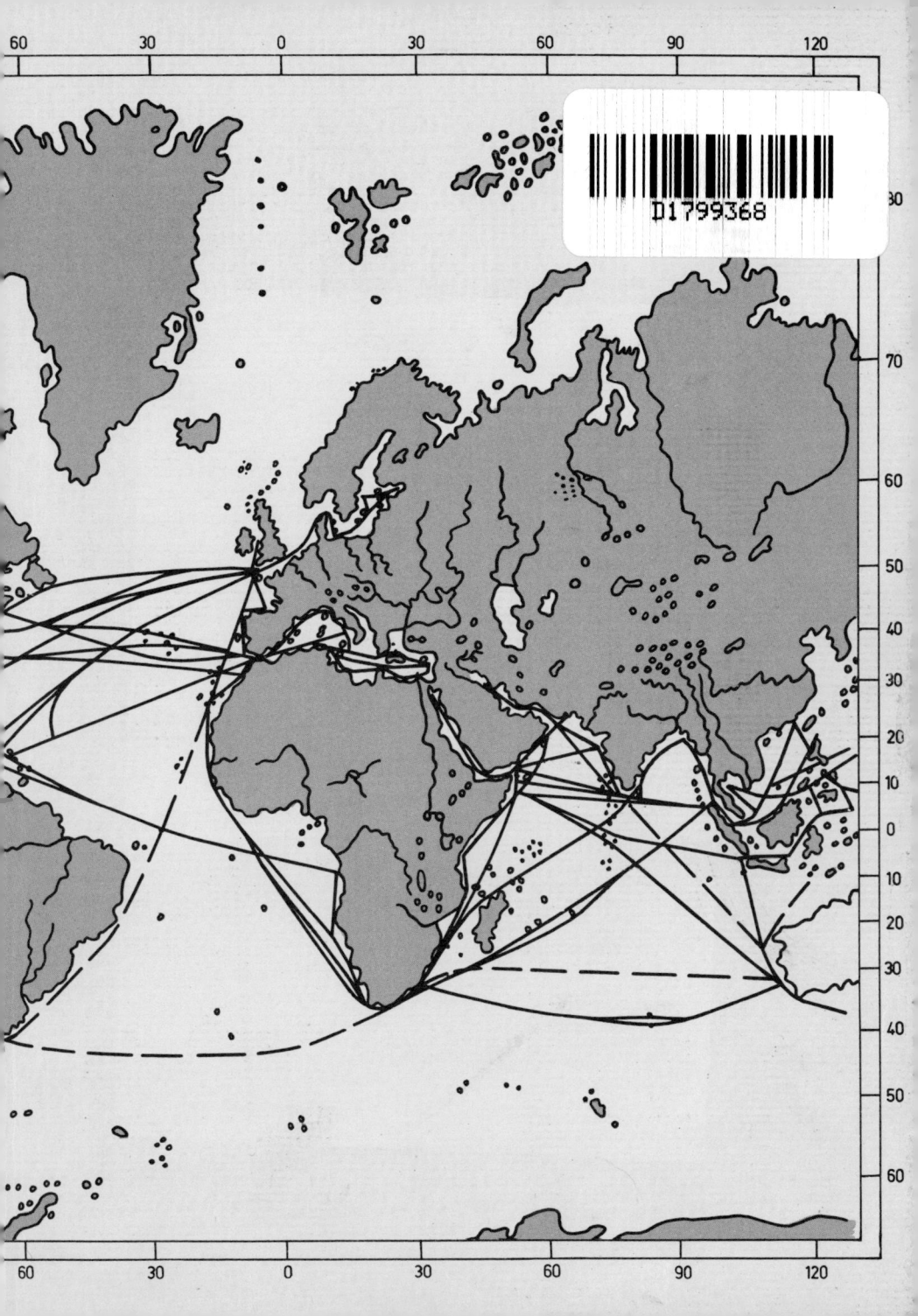

Field guide of
whales and dolphins

by

capt. W. F. J. Mörzer Bruyns

uitgeverij tor / n.v. uitgeverij v.h. c. a. mees
zieseniskade 14II
amsterdam
netherlands

Foreword

The mammals of the Order *Cetacea* – the whales, dolphins and porpoises – constitute one of the most fascinating and mysterious of all animal groups. The Blue Whale which is 25 ft long at birth and can grow to almost 100 ft long and 170 tons in weight, is not only the largest animal in the world, but so far as scientists know, the largest animal there has ever been in the world – about four times larger than any known dinosaur.

But it is not only size that has made this group of aquatic mammals so fascinating to mankind. A great deal of mystery still surrounds them. Some species, for example, are only known from skulls or skeletons and have never been seen in the flesh alive or dead.

And only quite recently has it been discovered that the group contains some of the most highly intelligent animals in the world.

The author has observed and studied the *Cetacea* during almost 40 years at sea, and has felt during that time the need for some kind of identification book. My own seafaring days have been limited to some 5 years during the second World War and to various liner voyages, but I too have felt the need for such a "field guide". I am therefore delighted that Captain Mörzer Bruyns has written this book, and that he should have invited me to write a foreword for it.

Captain W. F. J. Mörzer Bruyns is the twin brother of the distinguished zoologist and conservationist Professor Dr. M. F. Mörzer Bruyns. As a professional sailor the author must be regarded as an amateur zoologist, but it seems likely that no other living man has so much first-hand experience of so many species of Cetaceans in the wild state. Writing from this basis of personal experience Captain Mörzer Bruyns has made a number of new judgements about the relationships of the group which are somewhat unorthodox and may not be acceptable to all Cetaceologists, but which may well stimulate thought, discussion, and the further study which is so badly needed.

Although primarily intended as a popular guide for those travelling at sea, the book contains much hitherto unpublished material especially about the little known species. It is certain to stimulate a greater interest in these fascinating creatures and this is especially important at a time when pollution even of the great oceans has become a significant danger and when the problems of over-exploitation in the whaling industry have not yet been satisfactorily solved.

Whaling in its early days had, no doubt, an element of high adventure, but I believe that the judgement of history will record man's treatment of the great whales as one of the most shameful episodes in human progress.

The conservation of the living resources of our earth is only just beginning to receive the attention it must have if the human race itself is to survive. A heavy responsability rests on mankind which, with its explosions of technology and human population, has so drastically changed the biosphere in which all animals and plants must live.

That responsability seems to rest particularly heavily in respect of the Cetaceans because of their size, their unexpectedly high intelligence, their marvellous adaptation to their aquatic environment, the rarity of so many species and the immense amount which still remains to be discovered about them.

Peter Scott.

Slimbridge
February 1971

Chairman and 1st Vice President World Wildlife Fund
Peter Scott

Introduction

This book is meant to be a practical guide for all persons who are interested in whales, dolphins and porpoises: seamen, fishery people, biologists, naturalists and others.

Although the contents are based on careful observation and information from many zoological publications, this guide makes no claim to be a scientific work. The author is no professional zoologist. He, however, has sailed the oceans of the world for nearly 40 years, and has always had a keen interest in observing these animals in the field; an interest shared by many seafaring people, seamen as well as passengers. The sporting frolicking dolphins in front and alongside the ships and the enormous size and antics of the whales have the strong attraction of a glimpse of a strange and entirely different world.

The famous oceanaria in America and elsewhere exhibit the amazing intelligence and cooperation dolphins display in learning tricks and doing tests; this in latter years has given them worldwide fame through magazines, films and television.

The layman at sea sees many more different kinds, including some very rare ones, than most professional zoologists in Musea, Universities and Laboratories. With no complete guide available on the subject, it is impossible for him to recognize different species for himself, and incidentally for scientific purposes.

This guide is meant to help, by making identification easier to satisfy the observers' curiosity, and if they are keen enough, to provide information to professional zoology.

Throughout the last 18 years the author has made sketches and drawings of the animals at sea. Some of these took over 10 years to complete. Most of them could be traced back to species previously described in the quoted publications. Others could not and these perhaps are new species or forms which thus far have escaped the notice of science.

It is well known that zoology is still far from being able to publish an accurate and complete index. In this publication several problem species are described. It is not surprising, as many species of whales and dolphins are imperfectly known.

The author's sketches, transformed into watercolor pictures, are the illustrations of this guide. They are not perfect because the author is no artist. However, it proved impossible to find some one who could paint a good picture of a dolphin which he has never seen alive. The same kind of problem would be faced by an Eskimo artist if he should be asked to draw an illustration of a monkey.

The author has seen all the whales with the exception of the two rarest species, and nearly all the different species of dolphins and porpoises, alive at sea. With this background, and with the advice of biologists, he undertook to make this guide, because along with many others he felt a strong necessity for a complete (all known species of the world) and, as far as possible, accurate book.

In the existing publications, geographical distribution is treated rather vaguely, partly because those books were not written for that purpose, and partly because the distribution was not known.

The author is convinced after years of actively observing and recording that most dolphins and whales have a more restricted distribution than is generally supposed and he knows that the 100 fathom line (edge of the continental shelf), the seawater temperature and oceanic islands play an important part.

In the charts, the areas are indicated where they were seen by him and where they live according to information in official publications.

With fair regularity many kinds of whales and dolphins are stranded or swim to their death on coasts. These strandings unless frequent and in numbers, by no means always indicate that the animals live there. The author is more inclined to believe they beach themselves in many cases because they have strayed into a territory which is completely unfamiliar to them and does not give them a normal chance to survive.

As examples can serve: Sperm Whales and Bottlenosed Whales on the shallow and sandy Dutch coast, a Beaked Whale in the English Channel, and a Finner Whale Whale in the Gulf of Suez or the Malacca Straits.

The majority of whales and dolphins are afraid of a ship, some are indifferent and only a rather small part of them comes near to play in the bow or stern wave. The clear oceanic water quite often gives a good opportunity to see the whole animal even if it does not jump above the surface.

The texts are kept brief and as complete as possible so that observers may recognize a species at sea or stranded on a beach.

For practical reasons the author uses the classification into families, subfamilies, genera and species of: Victor B. Scheffer and Dale W. Rice 1963: a List of Marine Mammals of the World; Special Scientific Report, Fisheries nr. 431, United States Department of the Interior, Fish and Wildlife Service. The author presents this guide in the hope that it will be received by many as a welcome addition to the subject and that it will give the wonderful way of travelling by ship an added and fascinating interest.

Bussum
mei 1971

W. F. J. Mörzer Bruyns

Acknowledgements

The author likes to express his gratefulness and thanks for the encouragement, help, information and advice he received from the late Prof. Dr. E. J. Slijper of the Zoological Laboratory of the University of Amsterdam. He expresses also his thanks and appreciation for their help and support to Drs. W. L. van Utrecht, also of the Zoological Laboratory of Amsterdam, and to Drs. P. J. H. van Bree of the Zoological Museum of Amsterdam, and to Prof. Dr. M. F. Mörzer Bruyns of the Dept. for Nature Conservation of the University of Wageningen and Director of the State Institute for Nature Conservation Research (RIVON)* at Zeist, Holland, and more especially the Zoological Department, a.o. Miss M. D. M. Trommel, who made all the charts.

The RIVON gave valuable support to help this guide through the final stages, before publishing could be accomplished.

For the interest shown in his recordings and offering available information, publications and criticism, the author is much indebted to: Dr. F. C. Fraser of the British Museum of National History, London.

Prof. Dr. M. Nishiwaki of the Ocean Research Institute of the University of Tokyo, Japan.

Prof. Dr. H. Omura of the Whale Research Institute of Tokyo, Japan.

Dr. J. C. Moore of the Field Museum of National History, Chicago, USA, especially in the early stages.

Mr. H. van Deursen and Dr. G. van Gelder of the American Museum of National History, Central Park, New York, USA.

Dr. R. T. Orr of the California Academy of Sciences, San Francisco.

Dr. R. A. Falla, Mr. Charles McCann, Mr. J. Morland of the Dominion Museum, Wellington, New Zealand.

Mr. Logan, Mr. Frank and Mr. Bruce Robson formerly of Marineland Napier, New Zealand.

* Rivon is since 1969 part of the National Research Institute for Nature Management (RIN), Arnhem.

Valuable records of observations at sea came from Mr. K. Salwegter, Mr. F. J. Kerklaan and Mr. Adr. van Dam, of the Netherland Line in Amsterdam.
For presenting material or answering correspondence which was of great value in completing the book, the author expresses his thanks to the late Dr. Erna Mohr of Hamburg and to Mr. John Wallace of the Oceanarium at Durban and Dr. P. B. Best of the South African Museum (NH), Cape Town, South Africa. The author wishes to mention especially Mr. and Mrs. A. Prior for their beautiful Sealife Park at Waimanalo, Oahu, Hawaii, because many pictures taken there of the different rare species they collected, published in pamflets and newspapers, were a great help; and Mr. and Mrs. Arthur H. Spitzer, Consul of the Netherlands in Honolulu, Oahu, Hawaii, for keeping him informed about new additions in the Sealife Park and remarkable sightings in Hawaiian waters. The author wants to emphasize that although he gratefully acknowledges the help he received from zoologists from all over the world, those zoologists are in no way responsible for the views expressed and statements made in this book. The author is grateful and thankful to the publishers Miss Marianne Bettink and Miss Annemarie Behrens of "TOR" for their courage and enthusiasm which made the publishing of this guide possible.
Last but not least the author thanks the deckofficers and lookouts of the "Stoomvaart Maatschappij Nederland" at Amsterdam, 19 years under his command, who by their cooperation and frequent enthusiasm made all the recordings possible which enabled him to compile this book.
To them he wants to dedicate this guide.
The editing of this guide would not have been possible without considerable financial support of "De Nederlandse Stichting tot Internationale Natuurbescherming" (The Netherlands Foundation for International Nature Protection), at The Hague, Netherlands; and the "Oostersche Handel en Reederijen" at Amsterdam, Netherlands and of "The World Wildlife Fund", Morges, Switzerland, more especially of the Netherlands National Appeal "Het Wereld Natuur Fonds" at Zeist, Netherlands.
The author would appreciate receiving information or criticism which could be used to improve this guide or to make it more complete.
He is well aware of the many question marks which had to be omitted, and he apologizes for the mistakes he will have made, not being a zoologist himself, but working with information from zoological publications in his text, and combining them with his personal experiences. The manuscript was completed September 1968 and only minor alterations and corrections were added since.

The author

Whales

Whales are mammals, therefore warmblooded (31–37 °C), breathing air with lungs, through nostrils called the blowhole, which is situated on top of the head. Nature developed in whales living and feeding in water airways separated from the mouth and throat.

As already mentioned in the introduction, this guide gives no elaborate anatomical or historical descriptions and the reader whose interest in the subject goes deeper, is advised to read the following books:
Moby Dick by Herman Melville (1850) which is a classic on Sperm Whales.
(Giant Fishes), Whales and Dolphins by J. R. Norman and F. C. Fraser, London, 1948, which is more elaborate and explicit in many details, but less complete for many species and their distribution.
Whales by Prof. Dr. E. J. Slijper, Amsterdam 1962. A classical complete and interesting work on whales, their evolution, anatomy distribution and history. The 414 pages read as easily as a novel and reveal many secrets which nature developed to keep them warm or cool, how they "hold their breath" for a long time when diving to great depths, etc. Because it is a book about "whales" most of the smaller species, which make up 90 % of the group are only sporadically mentioned, and the book can not be used for identification.

Calves are born – tail first – usually one, occasionally two. They are nursed by the mother for up to 7 months in the larger species. The length at birth varies from $\frac{1}{3}$ to $\frac{1}{2}$ the length of the mother.
In whales the front legs have grown into pectoral fins, the flippers. Inside the skeleton of arm and hand is complete, but rigid; it can only move in the shoulder joint.
The rear legs have competely disappeared, only one or few loose bones in the body are left.
The tail end developed to a horizontal plane called the flukes.

12

This horizontal tail, against the vertical one of a fish, immediately distinguishes the dolphin from big fishes of the same size (tunny, swordfish and sharks). On the back a skinfold forms the dorsal fin.

The neck is short and the head can only slightly move. The eyes have eyelids, the earopenings are very small (diameter small knitting needle) and have no external shell.

Most whales can see and hear very well and, in addition, possess a sonar system, which, in selectivity, far surpasses similar instruments, so far produced by human intellect.

Man can hear "sounds" between 40 and 16.000 Hz (string bass to the chirping of grasshopper). The lowest human voice is 90 Hz.

A dog can hear to 35.000 Hz and a cat to 50.000 Hz. Dolphins can hear (receive) from 16 to 180.000 Hz and transmit to 150.000 Hz. The plankton-eating whales transmit from 30 to 400 Hz.

A "blind" dolphin (blindfolded on purpose or because the water is very muddy) can find his way unerringly through obstructions, has no trouble in detecting a hole in a net used to catch it, and can distinguish a morsel of cod from a piece of herring the same size.

In comparison to other mammals, they have more blood – it is needed to store oxygen. Most ribs are not connected to the breastbone. All bones are saturated with fat, which is considered to make them more flexible. Skullbones are not knitted hard together, but are cemented with cartilage. The brain is very well developed.

The heartbeat of whales is very low; for the large species perhaps only 4–6 beats per minute.

The division into Whales, Dolphins and Porpoises is a rather arbitrary one. A vague rule classes all animals over 30 feet as whales, between 6 and 30 feet as dolphins, and the smaller types, which in addition are very coastal and have blunt heads, as porpoises.

The 18 feet *Pseudorca crassidens,* however, is always called the False Killer Whale, whilst the 28 feet *Ziphius cavirostris* in some countries is called the Cuvier's Whale and in others the Cuvier's Dolphin.

There are some species which in most publications have no common name. For that reason the international Latin names are used in the texts and the index. In a number of cases common names are suggested.

In the alphabetical register they appear under both the Latin and the common name.

In the United States all but the bigger whales, are usually called porpoises, which adds to the confusion.

Division in orders, families, genera and species

Whales belong to the order of CETACEA, which is divided into two sub-orders:
1. *Odontoceti* (TOOTHED WHALES), with teeth. They eat fish, squid, etc., which are swallowed whole. These whales eat approximately 5 % of their body weight daily. Occasionally up to 20 %. The nostrils form *one* blowhole.
2. *Mysticeti* (BALEEN WHALES). The mouth is provided with "Baleens" or whalebone to sieve the food, which is plankton.
These whales eat 2 % of their body weight daily and can fast for months. The horny blades with a fringe along the inner edge "strain" the food from the seawater, like a sieve. The nostrils form two blowholes.
Of the 108 whales described in this guide, 98 belong to the toothed whales. Although several toothed whales reach lengths of up to 30 feet, only the Sperm Whale grows to what is considered "whale size".

Ten species are Baleen Whales, all but one of which reach whale size. The Blue Whale, reaching a length of 100 feet and weighing up to 170 tons is the largest animal alive and also the largest which has ever lived.

ODONTOCETI
The toothed whales are divided into four families:
1. RIVER DOLPHINS (*Platanistidae*)
   3 sub-families – 4 genera – 4 species.
2. OCEAN DOLPHINS (*Delphinidae*)
   2 sub-families, one with 2 genera and 2 species,
   the other with 20 genera and 64 species.
3. BEAKED WHALES (*Ziphiidae*)
   5 genera – 18 species.
4. SPERM WHALES (*Physeteridae*)
   2 genera – 3 species.

This division needs revising (after this book was accepted for publication a new division was made), but for this guide it is used for practical reasons. When science has eventually acquired sufficient knowledge, the family or group of OCEAN DOLPHINS will have more sub-families, genera and species. Five genera (*Phocoena, Neophocaena, Sotalia, Sousa* and *Orcaella*) are definitely not "Ocean Dolphins"; the genus *Stenella* is by "official" zoologists often stated to be a "chaos".

Recent strandings and publications of species which thus far were only known from skulls, indicate they were probably placed in the wrong genus.

It has already been suggested that the "Killer" group which the author has brought together in one chapter, should be classed as a separate family, the *Globicephalidae*.

Of the "Beaked Whales" very little is known, most of it only from few to very few stranded specimens.

The official division of the "List of Marine Mammals of the World" by Scheffer and Rice (1963), will be maintained in this book. However, where the accuracy is in doubt, the author has referred to a different opinion in the text.

MYSTICETI
The "Baleen Whales" are divided into 3 families:
1. BALEEN WHALES (*Balaenidae*)
   with 2 or 3 genera and 3 species.
2. GRAY WHALES (*Eschrichtidae*)
   with 1 genus and 2 species.
3. FIN WHALES or RORQUALS (*Balaenopteridae*)
   with 2 genera and 6 species.

Because of their commercial value these whales are better known than the toothed whales. This value has led to the near extinction of the first two families during the last century and the beginning of this one. Since they now enjoy an almost 100 % protection, there are hopeful signs of recovery.

During the last 9 years it has become quite obvious that the third family is now nearing destruction, and it is hoped that the combined efforts of the International Whaling Commission, The International Union for Conservation of Nature and Natural Resources (IUCN), the World Wildlife Fund and FAO, together with the governments of the countries participating in whaling, and the whaling interests themselves, can find a solution to save them.

Considering that a yearly revenue of £ 50,000,000 is at stake which because of the decreased stock, has already been reduced to less than half the amount mentioned above, the difficulties are understandable.

There is ample evidence that man has pursued the whale since the dawn of history. Around the 12th century, with the Basques, it became a trade, increasing in importance to a peak in the 17th century. The whales involved were Right Whales, and during the last part of the above period, Sperm Whales as well, because these could be approached and overtaken by rowing boats and did not sink when dead. Through lack of game the industry declined until, in 1865, the invention of the harpoon and the use of steam engines brought the Fin whales within reach. Thus far these had been too fast to overtake and they sank when dead.

A very interesting book, describing the ways in which mankind hunted whales all over the world from earliest known days is: "Follow the Whale" by Ivan T. Sanderson, 1956, Canada.

Conservation of whales and dolphins

While this guide was not initially written for this purpose, the author intends it as a contribution to further conservation and effective management of these interesting animals. Identification of species in the field is always one of the first steps to conservation.

Whales and dolphins, perhaps more than any other group of large animals, are common property of all nations. The management of populations of whales and dolphins is therefore an international matter.

This is especially so because the commercial use of these animals (whaling), of great importance for some countries, threatens several species with extinction.

In this chapter the author brings to the attention of the reader his views concerning the conservation of whales, based on his own experiences and on his contacts with nature conservation zoologists.

The first thing to do is to collect more exact data about all species, their numbers and their distribution in the different seasons. It will be necessary to catch specimens of several species for taxonomic studies. This guide shows clearly how necessary these surveys are, because so much is still unknown. It is even probable that some dolphins and whale species belong to the rarest animals living.

As soon as an outline is known about distribution and numbers, more systematical observations ought to be made to learn more about the ecology of the species, their requirements (temperature, food), the factors limiting the populations (including whaling) and their behaviour.

Regular counts have to be made in all seas and oceans and animals will have to be tagged or marked to get information about migration, age etc. Where and whenever possible species should be studied more closely (for instance, in oceanaria). The management of whale populations should be based on principles aiming at the conservation of all species. The author should like to quote here a statement of the International Union for Conservation of Nature and Natural Resources (I.U.C.N.) and especially its Survival Service Commission (S.S.C.) because these institutions, in his opinion, approach the problem exactly in the right way. The S.S.C. states:

"Proposals for the effective management of whales should be based on the results of these investigations, making concessions to economic criteria only in so far as optimum sustainable yields are not jeopardized, and should apply to all whaling nations.

Quotas for the numbers of whales to be harvested should include both pelagic and shore catches and should not exceed the required yields of the individual species and stocks. The "Blue Whale Unit" should, therefore, be abandoned in favour of a quota for each species.

Species which are shown to be seriously threatened by extinction, such as the blue, humpback and right whales, should be strictly protected, wherever they exist, until scientific investigation confirms that exploitation can recommence. Small numbers of protected species may be taken for scientific purposes, but this practice should be subject to rigorous, international control".

Whale populations ought to be managed according to these principles. Management in the opinion of the author is a matter of importance not only for nations with interests in whaling, but for all nations. National and international regulations are needed to provide for all species the protection necessary to prevent decrease of populations and extinction of species.

Each country, where one or more species of whales, dolphins or porpoises occur regularly within its territorial waters, ought to check its legal situation as to the conservation of these animals, to improve legal regulations if desirable, to prompt new legislation when necessary and to give full attention to the enforcement of the laws.

Many species will have to be protected completely all year round in their entire area. For other species an open season might be acceptable, but the author is convinced that for each species reasonably large reserve areas ought to be established, where they are not only free from persecution but can live undisturbed all year round.

As previously mentioned, more scientific research is needed. In the first place, research into the economically important species and into the species about which very little is known. The author is convinced that the "Gulland Plan", an United Nations agency dealing with the management and other problems of all whales, dolphins and porpoises, is the best way. This agency should co-operate with the International Whaling Commission and FAO for

18

the economically important species and be advised by I.U.C.N. together with UNESCO for the rare and scientifically important species. In the meantime, pending the establishment of this U.N. agency, the author sincerely hopes, that nature conservation departments and institutions in all countries concerned will be active in conserving whale populations within their own territorial waters and in supporting international activities. He also hopes that I.U.C.N. and World Wildlife Fund will be able to make progress in the international field as they have done before.

SUBORDER: ODONTOCETI: TOOTHED WHALES
(The numbers between brackets indicate the platenumbers)

FAMILY   *Platanistidae:* River Dolphins or Freshwater Dolphins

subfamily   *Platanistinae*
genus   *Platanista*
    – *Platanista gangetica*   Ganges D(olphin) (1)

subfamily   *Iniinae*
genus   *Inia*
    – *Inia geoffrensis*   Amazone D. or Boto (2)
genus   *Lipotes*
    – *Lipotes vexillifer*   Chinese Lake D. or
                         White Flag D. (3)
subfamily   *Pontoporinae*
genus   *Pontoporia*
    – *Pontoporia blainvillei*   La Plata D. (4)

FAMILY   *Delphinidae:* Ocean Dolphins

subfamily   *Monodontinae*
genus   *Delphinapterus*
    – *Delphinapterus leucas*   White Whale or Beluga (5)
genus   *Monodon*
    – *Monodon monoceros*   Narwhal (6)

subfamily   *Delphininae*
genus   *Phocoena*

	– *Phocoena phocoena*	Common Porpoise (7)
	– *P.p. vomerina*	id.
	– *P.p. relicta*	id.
	– *P. sinus*	Gulf of California P. (8)
	– *P. dioptrica*	Spectacled P. (9)
	– *P. spinnipinnis*	Burmeister P. (10)
genus	*Neophocaena*	
	– *Neophocaena phocaenoides*	Black Finless P. (11)
genus	*Phocoenoides*	
	– *Phocoenoides dalli*	Dall's P. (12)
	– *P. truei*	True's P. (13)
genus	*Cephalorhynchus*	
	– *Cephalorhynchus commersoni*	Commerson's Dolphin (14)
	– *C. eutropia*	White Bellied D. (15)
	– *C. heavisidei*	Heaviside D. (16)
	– *C. hectori*	Hector's D. (17)
	– *C. h. bicolor*	Pied Hector's D. (18)
genus	*Lagenorhynchus*	
	– *Lagenorhynchus albirostris*	White Beaked D. (19 a.b.)
	– *L. acutus*	White Sided D. (20)
	– *L. obliquidens*	Pac. White Sided D. (21 a.b.)
	– *L. cruciger*	Hourglass D. (22)
	– *L. obscurus*	Dusky D. (23 a-d)
genus	*Lagenodelphis* c.a.	
	– *Lagenodelphis hosei*	Serawak D. (–)
	– *Delphininae spec.*	Sth. China Sea D. (24)
	– *Delphininae spec.*	Malacca D. (24a)
genus	*Delphinus* c.a.	
	– *Delphinus delphis*	(Atl.) Common D. (25a)
	– *D. d. ponticus*	(East Med.) Common D. (25b)
	– *D. capensis*	Cape D. (25c)
	– *D. delphis*	Common D. (Pac.) (26)
	– *D. d. subspec. (?)*	(Austr.) Common D. (26a)
	– *D. or Stenella sp. (?)*	Agulhas D. (27)
	– *D. bairdi*	Black White D. (28)
	– *D. roseiventris*	Red Bellied D. (29)
	– *D. spec. (?)*	Java Sea D. (29a)
genus	*Stenella*	
	– *Stenella longirostris*	Long beaked D. (30) (31)

21

	– *S. microps*	Spinner D. (32)
	– *S. alope*	Bengal D. (33)
	– *S. longirostris subspec. (?)*	Arabian D. (34)
	– *S. longirostris subspec. (?)*	Galapagos D (35)
	– *S. caeruleoalba*	Bluewhite D. (36)
	– *S. euphrosyne*	Euphrosyne D. (37, 38, 39)
	– *S. spec. (?)*	Greek D. (40)
	– *S. plagiodon*	Gulfstr. Spotted D. (41)
	– *S. graffmani*	Gulf of Panama Spotted D. (42)
	– *S. frontalis*	Atlantic Spotted D. (43)
	– *S. attenuata*	Philippine D. (44)
	– *S. malayana*	Malay D. (45)
	– *S. malayana subspec. (?)*	Flores Sea D. (46)
	– *S. spec. (?)*	Senegal D. (47)
genus	– *Peponocephala spec. (?)*	Illigan D. (48)
genus	*Lissodelphis*	
	– *Lissodelphis borealis*	North. Right Whale D. (49)
	– *L. peroni*	South. Right Whale D. (50)
genus	*Steno*	
	– *Steno bredanensis*	Rough Toothed D. (51)
	– *S. perniger (?)*	Elliott's D. (52)
genus	*Sotalia*	
	– *Sotalia pallida*	Amazone River D. (53)
	(or. *S. fluviatilis*)	
	– *S. guianensis*	Guiana River D. (54)
genus	*Sousa (Sotalia)*	
	– *S. teuszi*	Camerun River D. (55)
	– *S. plumbea*	Leadcoloured D. (56)
	– *S. lentiginosa*	Speckled D. (57)
	– *S. borneensis*	Borneo White D. (58)
	– *S. chinensis*	Chinese White D. (59)
	– *S. spec. (?)*	Australian White D. (–)
genus	*Tursiops*	
	– *Tursiops truncatus truncatus*	Bottle-nosed D. (60, 60b)
	– *T.t. catalania*	Ind. Oc. Bottle-nosed D. (60a)
	– *T.t. gilli*	Pac. Bottle-nosed D. (61)
	– *T. nuuanu*	Little Bottle-nosed D. (62)
	– *T. aduncus*	Red Sea D. (63)
genus	*Grampus*	

	– Grampus griseus	Grampus or Risso's D. (64)
genus	*Globicephala*	
	– Globicephala melaena melaena	Pilot Whale (65)
	– G.m. edwardi	id. (65)
	– G. scammoni	id. (–)
	– G. macrorhyncha	id. (–)
genus	*Pseudorca*	
	– Pseudorca crassidens	False Killer (66)
genus	*Peponocephala*	
	– Peponocephala electra	Little Killer (67)
genus	*Feresa*	
	– Feresa attenuata	Pygmy Killer (68)
genus	*Orcinus*	
	– Orcinus orca	Killer Whale (69)
genus	*Unknown*	
	– Delphininae species	Alula Killer (70)
genus	*Orcaella*	
	– Orcaella brevirostris	Irrawaddi D. (71)

FAMILY	*Ziphiidae:* Beaked Whales	
genus	*Tasmacetus*	
	– Tasmacetus shepherdi	Tasman Wh(ale) (72)
genus	*Mesoplodon (and Diplodon)*	
	– Mesoplodon bidens	Sowerby's Wh. (73)
	– M. europaeus	Gulfstream Wh. (74)
	– M. mirus	Truc's Wh. (75)
	– M. grayi	Scamperdown Wh. (76)
	– M. stejnegeri	Sabre toothed Wh. (77)
	– M. carlhubbsi	Hubb's Wh. (77)
	– M. bowdoini	Andrew's Wh. (77)
	– M. ginkgodens	Ginkgo Wh. (78)
	– M. layardi	Straptoothed Wh. (79)
	– M. densirostris	Blainville's Wh. (80)
	– M. hectori	Hector's Wh. (–)
	– M. pacificus	Longman's Wh. (–)

genus	*Ziphius*	
	– *Ziphius cavirostris*	Cuvier's Wh. (81)
genus	*Berardius*	
	– *Berardius arnouxi*	Arnoux' Wh. (82)
	– *B. bairdi*	Baird's Wh. (83)
genus	*Hyperoodon*	
	– *Hyperoodon ampullatus*	Northern Bottle-nosed Wh. (84)
	– *H. planifrons*	Southern id. (–)

FAMILY	*Physeteridae:* Sperm Whales	
subfamily	*Physeterinae*	
genus	*Physeter*	
	– *Physeter catodon*	Sperm Whale (85)
subfamily	*Kogiinae*	
genus	*Kogia*	
	– *Kogia breviceps*	Pygmy Sperm Wh. (86)
	– *Kogia simus*	Dwarf Sperm Wh. (–)

SUBORDER: MYSTICETI: BALEEN WHALES

FAMILY	*Balaenidae:* Baleen Whales	
genus	*Balaena*	
	– *Balaena glacialis*	Black Right Wh. (87)
	– *B. mysticetus*	Greenland Wh. (88)
genus	*Caperea*	
	– *Caperea marginata*	Pygmy Right Wh. (89)

24

FAMILY   *Eschrichtidae:* Gray Whales

genus   *Eschrichtius*
         – *Eschrichtius gibbosus*   Gray Wh. (90)

FAMILY   *Balaenopteridae:* Finwhales or Rorquals

genus   *Balaenoptera*
         – *Balaenoptera acutorostrata*   Minke Wh. (91)
         – *B. borealis*   Sei Wh. (92)
         – *B. edeni*   Bryde's Wh. (93)
         – *B. physalus*   Fin Wh. (94)
         – *B. musculus*   Blue Wh. (95)
genus   *Megaptera*
         – *Megaptera novae angliae*   Humpback Wh. (96)

TOOTHED WHALES
Odontoceti

RIVER DOLPHINS OR FRESHWATER DOLPHINS (*Platanistidae*)

Sub-divided into 3 sub-families

Platanistinae	Genus *Platanista*
Iniinae	Genus *Inia*
	Genus *Lipotes*
Pontoporinae	Genus *Pontoporia*

Each genus has 1 species (nr. 1–4).

The dolphins of this family are fresh water dolphins – a name which the author prefers to river dolphins – with long narrow snouts. They live far upstream in the rivers mentioned in the text, except the *Pontoporia,* which lives in the lower reaches. They are slow moving and difficult to detect, although the *Inia* is also reported to be curious and sometimes noisy.
Their weight is less than that of salt water dolphins of the same length.

The most outstanding difference from all other dolphins, except the first sub-family of the "Ocean Dolphins", is, that the neck vertebrae are not grown together in one rigid structure. This enables them, like all other mammals, to turn their heads sideways as well as up and down.

genus	PLATANISTA Wagler, 1830 (= Susu Lesson, 1828). GANGES DOLPHIN; Susu, plate nr. 1.
species	*Platanista gangetica* Lebeck, 1801.
area	India and Pakistan. Rivers Indus, Ganges and Bramaputra from near the source to about 100 miles from the sea (Chart 1).
water temp.	From cold in upper reaches to 30° C.
length	6–8 feet (1.8–2.5 meter).
weight	Approx. 160 lbs (73 kg).
food	Fish and Shrimps.
teeth	4 × 29, sharp and conical, top and bottom interlocking, wearing off with age.
illustration	From life and illustrations. The colour is uniform grey, from very dark to light.
special features	The long thin beak (7″) is rarely seen in the muddy water. The round forehead followed by the back as far as the ridge-like fin, comes smoothly and quietly to the surface. The dolphin then arches back and tail and dives steeply out of sight. In turbulent waters, around the dock entrances in Calcutta harbour, they occasionally jump above the surface and as true dolphins return head first.
speed	Cruising 2 to 4 knots, capable of more, they easily stem a tide of 7 to 8 knots.
breath	Every 30 to 45 seconds.
schools	No pronounced schools, often single, sometimes 4 to 6 together spread out and each surfacing individually.
biotope	Muddy waters and strong currents.
immatures	Calves born between April and July. Eight to nine months after mating.
notes	Seafarers will meet this dolphin on the Hoogly River to Calcutta, where during the summer, it comes downstream to just below Budgebudge. Also at Chalna anchorage Pussur River. The Susu is occasionally caught by the inhabitants along the river, who use the meat, blubber and oil. This dolphin is completely blind, the eye has no lens.

SUBFAMILY INIINAE

genus	INIA d'Orbigny, 1834.
	AMAZONE DOLPHIN, local name BOTO, plate nr. 2.
species	*Inia geoffrensis* (Blainville, 1818).
area	Widespread in upper reaches of the Amazone and Orinoco rivers, with tributaries (Brazil, Peru, Bolivia, Guiana and Venezuela); down the Amazon river to near the mouth (Chart 1).
water temp.	Above 25° C.
length	7 to 9 feet (2.10–2.75 meter).
weight	150–240 lbs; 22 (♀) to 27 (♂) lbs/ft (68–110 kg).
food	Fish of 6–12″, which they stir up from the bottom and catch near the surface. Their diet includes the much feared Piranyas.
teeth	4 × 25 or 28, conical and with rough crowns.
illustration	From photographs and cast. Usually dark to black above, lighter to pink underneath, sometimes all light grey to flesh colour. The colour probably becomes lighter with age. Captive animals of light colour in clear water and exposed to the sun become darker.
special features	The long thin beak rarely shows. The animal surfaces in horizontal position, exposing the blowhole, back and dorsal ridge above the surface, then arches for the next dive. When playing they sometimes jump out of the water to a height of four feet. They are curious and attracted by both noise and light on the surface. With their very small eyes they can see very well, also above the surface. The ear opening is large.
speed	Cruising two, maximum ten knots when swimming along the surface.
breath	Every thirty to forty five seconds with a maximum of 112 seconds dive. The blast is reported to be six feet high and very noisy. The local name is an imitation of the sound produced. The blowhole is a rectangular slit.
schools	Small, often in pairs and in company with the Amazone River Dolphin or Bufeo (*Sotalia pallida* or *S. fluviatilis*) (nr. 53), but twice as numerous.
biotope	Midstream small and fast rivers, also in lagoons, but rarely in inundated forests, comes to 1500 miles from the sea.

28

	Four to five per square mile.
immatures	No special time has been recorded for the birth of calves.
notes	They are held in high esteem by the Indians living along the river banks, who believe that killing them will cause bad luck, and that using the oil for light will blind the person who looks at it. More civilized tribes, however, hunt and use it. Other tribes are reported to use the dolphin's hunting habits to catch the pursued fish in their nets.
	There are two pairs in captivity in zoos in America. The author is indebted to Dr. Erna Mohr for much of this information, which she published in 1963.

genus	LIPOTES Miller, 1918.
	CHINESE LAKE DOLPHIN or WHITEFLAG DOLPHIN; Pei chi, plate nr. 3.
species	*Lipotes vexillifer* Miller, 1918.
area	Tung Ting Lake, approximately six hundred miles up the Yangtse River (Chart 1).
water temp.	30° C.
length	7–8 feet (2.1–2.5 m).
weight	Approximately 180 lbs (81 kg).
food	Fish (Catfish).
teeth	4 × 33 to 36, with rough crowns.
illustration	From cast. A photograph in the 1918 paper by Miller gives the impression of a more slender animal. The colour is grey to blue-grey above and white underneath.
special features	The blowhole is situated on the left side of the head and is rectangular in shape. The snout is 12″ (30 cm) long.
speed	Not recorded. Probably cruising two to four knots.
breath	Not recorded. Probably every 30 to 45 seconds.
schools	From three or four to ten or twelve.
biotope	Living only in the above mentioned lake and surrounding waters. In the dry season this area becomes very shallow and the dolphins herd together in the remaining pools.
immatures	No records.
notes	This dolphin is not killed on purpose by the surrounding inhabitants. There is an old legend attached to it which connects it with the drowning of a Chinese princess. However, when it is accidently caught, meat and blubber are used.

genus	PONTOPORIA Gray, 1846. (Stenodelphis d'Orbigny & Gervais, 1847). LA PLATA DOLPHIN; Franciscana, plate nr. 4.
species	*Pontoporia blainvillei* d'Orbigny et Gervais, 1844.
area	Lower reaches La Plata River, Argentinia (Chart 1).
water temp.	15–25° C.
length	Male: 5 feet (1.5 m); Female: 5'8" (1.7 m); Calves: 18" (0.46 m).
weight	Male: 70 lbs (32kg); Female: 90 lbs (41 kg); Calves: 15 lbs (7 kg).
food	Fish (Mullet).
teeth	4 $\times$ 48 to 50, fine and pointed like toothpicks (6,5 $\times$ 1,5 mm).
illustration	From cast and photographs. The colour is pale brown above and lighter below.
special features	Their small size and quiet ways make them difficult to detect in the muddy waters. They come to the bows and alongside of small fishing vessels.
speed	Not recorded; probably 2 to 4 knots.
breath	Not recorded; probably once or twice every minute.
schools	Small.
biotope	In the La Plata River delta. During the winter they are rarely seen and it seems possible that there is a migration northwards along the coast.
immatures	Calves are probably born in spring and summer.
notes	Little is known about these dolphins; but they are caught and used by fishermen in that area.

The title "Ocean Dolphin" is misleading and as far as "locality" is meant this family could be better divided into five parts:

OCEAN DOLPHINS: For the pelagic species.

MARINE DOLPHINS: For those who normally come as far as 150 to 200 miles from land.

COASTAL DOLPHINS: For the species who live coast wise; normally not further off shore than approximately 50 miles, and on the continental shelf.

ESTUARY or DELTA DOLPHINS: Whose favourite locality is shallow coasts and river entrances. Normally not seen further than two miles off shore.

RIVER DOLPHINS: For the three species or sub-species which frequent rivers upstream. Because this "locality-group" can be mistaken for the family *Platanistidae,* it might be better to refer to those as FRESH WATER DOLPHINS, which is what they really are and to restrict RIVER DOLPHINS for the *Delphinidae*-group.

Delphinidae are divided into two sub-families.

MONODONTINAE   Genus *Delphinapterus* – one species.
         Genus *Monodon*   – one species.
         Both species are arctic in distribution and have "movable" neck vertebrae in common with the Fresh Water Dolphins.

DELPHININAE   With 20 genera and ± 70 species.
         In this second sub-family, practically all dolphins are brought under one heading without a sub-division, except in genera.
         This simplification has been till now sufficient, but is not so practical in a guide to be used for identification.
         For the benefit of the observer, the author has made a special sub-division in characteristic groups of one genus, several genera, or part of a genus. For each group a short introduction is given.

genus	DELPHINAPTERUS Lacépède, 1804.
	WHITE WHALE or BELUGA, plate nr. 5.
species	*Delphinapterus leucas* (Pallas, 1776).
area	Arctic coasts of Europe, Canada, USA and Eastern Siberia (Chart 1).
	The populations of the Okhotsk Sea and Eastern Siberia are sometimes classed as sub-species:
	D.1. dorofeevi Barabash & Klumov 1935
	and those around the Kara and White Seas:
	D.1. marisalbi Ostroumov 1935 (= D.1. freimani Klumov).
sea temperature	0–5° C.
length	12–14 feet, occasionally to 18 feet (3.6–4.3 m; 5.5 m).
weight	2400 lbs (170 lbs/ft); (1100 kg).
food	Fish, Cuttlefish and Prawns.
teeth	10 in each upper jaw, 8 in each lower jaw. Diameter 0,8″ (2 cm).
illustration	From photograph and cast.
	Colour cream to milk-white above the age of 5 years. Calves are dark grey-brown, become lighter with age via a marbled stage when 8 feet long, to the whiteness of the adult.
special features	Swims slowly with undulating movements and shows very little of the head and back when coming to the surface for air. No dorsal fin.
speed	Maximum 6 knots.
breath	Not recorded, probably once or twice a minute.
schools	5 to 10, often swimming in single file.
biotope	A "Coastal, Estuary and River dolphin" in the Arctic. Recorded 700 miles up the Yukon River and on the St. Lawrence River as far as Quebec.
immatures	Born from April to June, 12 months after pairing. Length 5 feet (1.5 m). Are nursed for eight months.
notes	Immatures sometimes stray south as far as the British Isles, the Northsea and the Baltic, occasionally to the Bay of Biscay. During the last century, they were intensively hunted by the Norwegians for their hides and oil.
	A yearly catch of 2000 around Spitsbergen is now reduced to 300 and even as few as 20.

The species has recently been caught alive in St. Lawrence River by an ancient method using a fish-trap of wooden poles, where they fall dry at low water. The purpose of this venture is to sell them to different sea-aquaria.

1877–1878 there was twice one in a Westminster Aquarium (London) but both died after a few days of captivity. The author believes there was one in an Aquarium at Coney Island, New York, in 1955. In 1970 animals are said to live in the Zoo of Duisburg (Germany). Several "marinelands" obtained specimens since 1968.

During the summer 1966 one adult spent several weeks on the Rhine River far into Germany.

All attemps to capture it failed and it found its way back to the North Sea.

The heartbeat of this whale was registered as between 12 and 20, the animal was wounded by a harpoon (Alaska).

genus	MONODON Linnaeus, 1758.
	NARWHAL, plate nr. 6.
species	*Monodon monoceros* Linnaeus, 1758.
area	North Atlantic Arctic (Chart 1).
sea temperature	0–5° C.
length	13–18 feet (4.0–5.5 m).
weight	Approximately 1 ton.
food	Squid, Fish (preferable Flounders) and Shrimp.
teeth	One in each half of the upper jaw. In males one or very rarely both grow out to considerable length (9 feet), forming a tusk twisted in a right spiral. In females the teeth are usually not developed or form a small tusk.
illustration	From plates, general colour greyish-white with grey to black spots on the back, becoming lighter on the flanks and underside.
	The immatures are dark grey and almost identical to the calves of the Beluga.
special features	The marbled colouring, absence of dorsal fin and for the males the tusk. Very slow.
	No dorsal fin but a ridge 2–3 feet long (40–60 cm); 1–2" high (2.5–5 cm).

speed	Maximum 4 knots.
breath	Not recorded, probably once or twice a minute.
schools	6–10 animals.
biotope	Marine, sometimes coastal. Always near or in open spaces between ice fields.
immatures	Calves are probably born during the summer and have a length of 5 feet (1.5 m).
notes	This dolphin lives far outside all normal shipping routes. It is probably the "Unicorn" of old legends.

When harpooned, it is known to dive vertically to a depth of 200 fathoms, then comes to the surface in the same place. There sometimes are fights to the death between this species and the Walrus for "breathing holes" in the pack-ice.

The tail in older specimens is modified.

The blunt point of the thick flukes is turned forward. The flukes are not needed for speed, but this dolphin which sleeps in the breathing holes, perhaps uses it to swim backwards.

If the long tusk is used to probe the sea bottom, which is not certain, they might also have to "back out" of there sometimes.

They can make a whistling sound, and the mothers call their calves with a low tone.

SUBFAMILY DELPHININAE

The groups are:

I	PORPOISES	Genus *Phocoena*	4 species,
		Genus *Neophocaena*	1 species
		Genus *Phocoenoides*	2 species
II	SUB ANTARCTIC PORPOISES	Genus *Cephalorhynchus*	4 species
			1 sub-species
III	PLOUGHSHARE-HEADED DOLPHINS		
		Genus *Lagenorhynchus*	5 species
IV	S.E. ASIA DOLPHINS	Genus *Lagenodelphis*	1 species
		Delphininae. Spec.	1 or 2 species
V	COMMON DOLPHINS	Genus *Delphinus*	3 species, 2 or 3 sub-species

VI	LONGBEAKED DOLPHINS	Genus *Stenella*	3 species, perhaps 3 sub-species
VII	HARNESSED DOLPHINS	Genus *Stenella*	2 species, perhaps 2 sub-species
VIII	SPOTTED DOLPHINS	Genus *Stenella*	4 species
IX	NARROW-SNOUTED DOLPHINS	Genus *Stenella*	2 species, perhaps 3
X	UNIDENTIFIED DOLPHINS	Genus *?*	2 illustrations
XI	RIGHT WHALE DOLPHINS	Genus *Lissodelphis*	2 species
XIA	ROUGH TOOTHED DOLPHINS	Genus *Steno*	2 species
XII	RIDGEBACKED DOLPHINS	Genus *Sousa*	5 species perhaps 6
		Genus *Sotalia*	2 species
XIII	BOTTLENOSED DOLPHINS	Genus *Tursiops*	3 species, 1 sub-species
XIV	BLUNTHEADED DOLPHINS	Genus *Grampus*	1 species
		Genus *Globicephala*	1 species, 1 sub-species
		Genus *Feresa*	1 species
		Genus *Pseudorca*	1 species
		Genus *Orcinus*	1 species
		Genus *Peponocephala*	1 species
		Genus *?*	1 unknown
XV	IRRAWADI DOLPHIN	Genus *Orcaella*	1 species, 1 sub-species

GROUP I   PORPOISES

Porpoises are small, blunt headed, coastal cold water dolphins.
Genus Phocoena and *Neophocaena* are slender, and lighter than a dolphin of the same size. They frequent shallow inshore waters and occasionally ascend rivers.
The species of the *Genus Phocoenoides* are heavier and can be met with further out to sea.
The author reckons as "porpoises" the species of the genera:
Phocoena (nrs. 7, 8, 9 and 10), *Neophocaena* nr. 11 and *Phocoenoides* nrs. 12 and 13.

genus	PHOCOENA G. Cuvier, 1817.
	COMMON PORPOISE or HARBOUR PORPOISE, plate nr. 7.
	including: *Ph. ph. vomerina* Gill, 1865 North Pacific
	Ph. ph. relicta Abel, 1905 Black Sea
	which, by some authorities, are regarded as different races. See notes.
species	*Phocoena phocoena* (Linnaeus, 1758).
area	Temperate and sub-arctic coasts of Northern Hemisphere (Chart 2).
sea temperature	Cold, preferably around 10° C.
length	5–6 feet (1.5–1.8 m).
weight	100–165 lbs; 22 lbs/ft (45–75 kg).
food	Fish, Squid and sometimes Shrimp (Herring, Whiting, Sole).
teeth	4 × 23, sometimes to 27, spade shaped, diameter 0,1″ (2.5 mm).
illustration	From life, cast and photographs.
special features	It prefers such shallow waters and normally stays so close inshore that ocean-going ships rarely see it except in harbours and passages between islands (Denmark, Scotland, Long Island Sound USA). It has a triangular dorsal fin and yachting people meet it frequently because it comes to the bow of small boats and plays alongside.
speed	Maximum about 8 knots.
breath	Every 15 seconds.

schools	50–100.
biotope	Shallow coastal waters, inland bays and sometimes up rivers. Generally depths less than 10 fathoms.
immatures	Pairing in summer, calves born after 8–10 months, usually from March to July. They are half the length of the mother and weigh 7 lb (3 kg).
notes	The meat of porpoises was eaten throughout the middle ages until the 17th century; also at the English Court. Whales and Dolphins are "Crown Property" in England, and nobody is allowed to hunt or kill them. The animals inhabiting the coast of N.W. Africa are described as bigger and lighter in colour, and it is suggested that these are "older". First hand information from a Bulgarian "dolphin harpo-neer" indicates that the Black Sea porpoise only resides in fresh or brackish water of the sea of Asov, and that these dolphins have a longer snout. The name is "Asovski mitkur" and the length is about 4 feet, the weight approximately 60 kg.

GULF OF CALIFORNIA PORPOISE, plate nr. 8.

species	*Phocoena sinus* Norris & MacFarland, 1958.
area	Gulf of California – Mexico (Chart 2).
sea temperature	Above 20° C.
length	5 feet (1.5 m).
weight	110 lbs (50 kg).
food	Fish (*totoaba*) and probably Shrimps.
teeth	Not recorded, but probably the same as Common Porpoise.
illustration	From description. The colour is lead-grey above to white below. No dark stripe from beak to flipper.
special features	Contrary to the playful habits of the Common Porpoise, it is wary and difficult to observe. They follow erratic courses under water and surface in unexpected places.
speed	Not recorded, probably maximum about 8 knots.
breath	Not recorded, probably every 15 seconds to one minute.
schools	2 to 6.
biotope	Shallow coastal waters and harbours in the Guf of Cali-fornia, etc. Depths less than 10 fathoms.
immatures	Calves are born in May and June. The length is 2 feet.
notes	No specimen has been acquired for scientific study, al-

though local fishermen know it well, and sometimes catch it in their fishing and shrimp nets.

SPECTACLED PORPOISE, plate nr. 9.

species	*Phocoena dioptrica* (Lahille, 1912).
area	Patagonia to South Georgia (Chart 2).
sea temperature	5–18° C.
length	5 feet (1.5 m).
weight	Approximately 120 lbs (54 kg).
food	Fish and probably Cuttlefish.
teeth	4 × 19 to 21. Very small and spade shaped.
illustrated	From description and other illustrations. The dark line from beak to flipper is not always present.
special features	The striking black and white colouring. The similarly coloured Southern Right Whale Dolphin nr. 50 is slightly larger and has no dorsal fin.
speed	Probably maximum 8 knots.
schools	Not recorded.
biotope	Cold inshore waters of the Falkland Current.
immatures	Not recorded.

BURMEISTER PORPOISE, plate nr. 10.

species	*Phocoena spinnipinnis* (Burmeister, 1865).
area	East and west coasts of South America (Chart 2).
sea temperature	0–10° C.
length	5 feet (1.5 m).
weight	Approximately 120 lbs (54 kg).
food	Fish and probably Cuttlefish.
teeth	4 × 16 to 17.
illustration	From description.
special features	The unusual shape of the dorsal fin.
speed	Probably maximum 8 knots.
breath	Probably every 5 to 15 seconds.
schools	No records.
biotope	Colder inshore waters, probably depths less than 10 fathoms.
immatures	No records.
notes	This is a rare porpoise, very little is known about it, the author has not seen it in Calao, and several of his colleagues with experience of intensive coastal navigation between this port and Valparaiso have never noticed it either.

genus	NEOPHOCAENA Palmer, 1899.
	BLACK FINLESS PORPOISE; Nomeno Juo, plate nr. 11.
species	*Neophocaena phocaenoides* (G. Cuvier, 1829).
area	East Asia, Yangste Kiang to 1000 miles up river. East and South Coast Honshu to Shimonoseki, also in Bombay harbour (India) and has sporadically been reported between these extremes (Chart 2).
sea temperature	5–15° C, occasionally to 25° C.
length	4'6"–5' (1.4–1.5 m).
weight	100 lbs (45 kg).
food	Fish, Cuttlefish and Shrimps.
teeth	4 × 15 to 19 spade shaped.
illustration	From life (Bombay). Black to very dark grey, lighter areas between the flippers and around the anus. They are reported to have pink eyes in North East Asia.
special features	Small size, no dorsal fin but a sharp ridge. Anterior to the ridge a slight dent on the back. They are very lively, swim just under the surface with quick sudden movements right and left and sometimes in circles. The blowhole just touches the surface for breathing and they do not jump.
speed	2 to 3 knots, faster in spurts.
breath	3–4 times in quick succession every 7–13 seconds, then a dive lasting from 45–75 seconds.
schools	In pairs to about six.
biotope	Estuary dolphin, shallow and muddy water, fresh and brackish, depths less than ten fathoms.
immatures	Calves are born in August.
notes	In the home waters in North East Asia, the temperature of the sea water ranges from 5–10° C in January and February; 10–15° C in May; and up to 20–25° C from July to October, which is the largest seasonal change in water temperature known by the author. This classifies the dolphin as a cold water species and in the warmer water of the Kuro Shio passing along the South coast of Japan it is rarely seen. More frequently in the large bays and the Inland Sea, where during winter the water temperature ranges between 4 and 6° C. The population in Bombay Harbour is purely tropical with

sea water temperatures ranging between 27 and 30° C. They are difficult to observe but seem to be nearly always present along the Ballantine Pier at all tides. Their progress in a deep dive is about 50 to 100 yards.

Existing illustrations give the porpoise a fuller, squarer forehead than the ploughshare shape in this book. Perhaps the N.E. Asia animals are different from the S. Asia population. Special features in the skulls conform this.

genus	PHOCOENOIDES Andrews, 1911.
	DALL'S PORPOISE, plate nr. 12.
species	*Phocoenoides dalli* (True, 1885).
area	North Pacific Ocean, temperate and sub-arctic, south to Japan (39° N) and California (34° N) (Chart 1).
sea temperature	0–15° C, occasionally 20° C.
length	5′–6′6″ (1.5–2.0 m).
weight	200–300 lbs (average 43 lbs/ft) (90–136 kg).
food	Squid and sometimes Fish, mostly caught at considerable depth.
teeth	Upper jaw 2 × 23, lower jaw 2 × 27, very small.
illustration	From life and photographs.
special features	The white field on the flank immediately identifies the species. They approach the surface at full speed in a horizontal position, producing a conspicuous "bow wave" with the head and dorsal fin. They are not afraid of ships and if they are not too fast, come to the bow and play, also swim alongside at a distance of $\frac{1}{4}$ to $\frac{1}{3}$ nautical miles. They do not jump.
speed	12 knots.
breath	Every 17 seconds, or twice in 28 seconds. Can stay under for 5 minutes.
schools	From 2–6, but sometimes from 20–40.
biotope	Coastal to oceanic but in deep water. The majority were seen 10 to 5 miles off the coast in depths of 100 to 500 and 1000 fathoms.
immatures	Calves are probably born in the north during the summer (they were three-quarter length during February off Los Angeles).
notes	When caught, they are reported to put up a terrific fight, then

die. So far it has not been possible to keep them alive in one of the oceanaria. They undoubtedly migrate; north in the spring, south in the autumn; but in the approaches to Los Angeles can be seen every month of the year, During July they were seen in that area in company with Humpback Whales.

This species and the next have 30 more vertebrae than the Common Porpoise.

	TRUE'S PORPOISE, plate nr. 13.
species	*Phocoenoides truei* Andrews, 1911.
area	Japanese waters, between 30° and 45° N, as far as 200 to 300 miles offshore (Chart 1).
sea temperature	0–15° C, occasionally 20° C.
length	5 feet (1.5 m).
weight	Approximately 220 lbs (100 kg).
food	Squid.
teeth	Upper jaw 2 × 19; lower jaw 2 × 22; very small.
illustration	From description. Except for the size of the white area on the flanks, this dolphin is identical to the Dall's Porpoise (nr. 12).
special features	The white area on the flanks is about double the size of that of the Dall's Porpoise. In Northern Japan the territories of these two species overlap.
speed	Approximately 12 knots.
breath	Probably every 15–17 seconds.
schools	2–6, and more.
biotope	Japanese waters, coastal and marine, depths over 100 fathoms.
immatures	Not recorded, but probably the same as in Dall's Porpoise.
notes	During two visits to Japan, south coasts of Kyushu, Shikoku and Honshu, in February (sea temperature 17° C) and June (sea temperature 24° C), only one school of probably these dolphins was seen ENE of Cape Shiono.

Some zoologists have considered the Dall's and True's Porpoise to be one species. The fact that schools of both are seen simultaneously but never mix, is in favour of the opinion that they are separate species.

GROUP II   SUB ANTARCTIC PORPOISES
(SMALL SOUTHERN DOLPHINS)

A group of small cold-water dolphins, in size and shape more like porpoises than what are generally considered dolphins.

They belong to one genus: *Cephalorhynchus,* Gray 1846.

The forehead (melon) is less pronounced and in consequence the shape of the head more flat and triangular.

They do not live near the normal shipping routes, observations are rare and little is known about them.

Their range is mainly sub-antarctic in areas with sea water temperature of 12° C and lower, and from what is known, very coastal in small selected area's.

Four species are recognized, one of them with a sub-species. The colouring is black and white or white and black, which with the small size and blunt head makes identification not difficult.

genus	CEPHALORHYNCHUS Gray, 1846.
	COMMERSON'S DOLPHIN, plate nr. 14.
species	*Cephalorhynchus commersoni* (Lacépède, 1804).
area	Sub-antarctic-Patagonia to Falkland Islands, also reported from the Kerguelen in the Indian Ocean (Chart 1).
sea temperature	5–10° C.
length	4–5 feet (1.4–1.5 m).
weight	Not recorded. Approximately 120 lbs (54 kg).
food	Cuttlefish and Shrimps and Krill.
teeth	4×29 or 30.
illustration	From description, illustrations and photo of related species.
special features	Striking black-white colour pattern. Small and fat.
speed	Not recorded. Maximum probably 6–8 knots.
breath	Not recorded, probably every 5–10 seconds.
schools	Single to 3.
biotope	They seem to prefer fjord-like bays and channels as are typical of Tierra del Fuego and the Falkland Islands. Inshore waters.
immatures	No records.
notes	It is quite possible that these dolphins also occur around the antarctic islands between the two groups mentioned above.

WHITE BELLIED DOLPHIN or CHILIAN DOLPHIN, plate nr. 15.

species	*Cephalorhynchus eutropia* (Gray, 1849), (= *C. albiventris*, Perez 1896).
area	Coast of Chili (Chart 1).
sea temperature	5–15° C.
length	4 feet (1.4 m).
weight	Approximately 120 lbs (54 kg).
food	Cuttlefish and Shrimps.
teeth	Upper jaw 2 × 28/30, lower jaw 2 × 28/31).
illustration	From description.
special features	The white areas on the belly are not noticeable in the field.
speed	Not recorded, maximum probably 6–8 knots.
breath	Not recorded, probably every 5–10 seconds.
schools	No records.
biotope	Inshore waters.
immatures	No records.
notes	Very little is known about this dolphin, it resembles the Commerson Dolphin but is much darker. It is very rare and probably does not range north of Chiloe Island.
	Another name is "Black Dolphin", which is as little help as identification as "White Bellied Dolphin".
	The author likes to suggest the name "Chilian Dolphin".

HEAVISIDE DOLPHIN, plate nr. 16.

species	*Cephalorhynchus heavisidei* (Gray, 1828).
area	Around Cape of Good Hope (Chart 1).
sea temperature	12–18° C.
length	4 feet (1.2 m).
weight	Approximately 100 lbs (45 kg).
food	Cuttlefish and Shrimps.
teeth	4 × 25 to 30, small and pointed.
illustration	From description. (The original sketch was made in 1956).
special features	Very small size, and the conspicuous white areas in the black of the flanks and tailstock.
speed	No record.
breath	No record, probably every 5–10 seconds.
schools	No records, probably small.
biotope	Probably cold inshore waters.
immatures	No records, only known from 5 specimens.

notes	These animals are cold-water dolphins, preferring a fjord-like coast. This species lives probably inshore north of Cape Town where the coast is type-locality. The temperature of the sea water varies from 12–15° C.

These animals are cold-water dolphins, preferring a fjord-like coast. This species lives probably inshore north of Cape Town where the coast is type-locality. The temperature of the sea water varies from 12–15° C.

This dolphin was only known from 2 specimens.

The first was captured in 1856, the second in 1965 both at Green Point, Cape Peninsula.

Three more specimens were caught further north along this coast in 1969.

HECTOR'S DOLPHIN, plate nr. 17 and
PIED HECTOR'S DOLPHIN, plate nr. 18.

species	*Cephalorhynchus hectori* (Van Beneden, 1881).
subspecies	*C.h. bicolor* Oliver, 1946 (= *C. albifrons* True 1899).
area	Restricted distribution in New Zealand coastal waters (Chart 1).
sea temperature	5 to 20° C.
length	$4\frac{1}{2}$–5 feet (1.4–1.6 m).
weight	Approximately 100–120 lbs (45–55 kg).
food	Small Fish, Cuttlefish and Shrimps.
teeth	4×30 to 32.
illustration	From life and photographs.
special features	A small quiet dolphin which shows little when it surfaces to blow, except the dark and peculiar dorsal fin. For this reason the original description of Van Beneden is of a dark dolphin.

Oliver described the bicolor in 1946. The author and everybody he met, who knew this dolphin, have only seen the bicolor variety. The eastcoast population is grey, the skin is covered with many tiny dark spots, the westcoast population seems to be pure white.

There is a great similarity with the Commerson's Dolphin, but with more white in the head and tailstock.

They swim in a rather undulating fashion and the author met several people who thought they were seals. Jumping is rare.

speed	Cruising no more than 2 to 3 knots, maximum estimated 6 to 8 knots. They regularly escort fishing vessels in and out of port and also try to approach bigger ships. They can re-

	main stationary on the surface.
breath	At sea they blow normally 3 or 4 times every 10 seconds between deep dives lasting 1 to 3 minutes, in the harbours the intervals are 4 tot 7 seconds. Progress during a deep dive is 100 to 200 meters.
schools	Small schools of 2 and 3, these schools 2 to 5 miles apart are normal. Large concentrations of several hundred have been reported off River bars (Rakaia River and Kawhia Harbour). Distance between individuals is 1 to 3 feet.
biotope	Shallow (green) inshore waters in depths from 3 feet to 30 fath. frequenting river bars and entrances to deep bays (Banks Peninsula) and harbours. Maximum distance off shore 2, occasionally 5 miles.
migration	There seems to be a northward migration along the west coast where 200 were observed off Kawhia Harbour (North Island) from December to January for 3 successive years.
immatures	Calves are born in December and January and are also "pied".
notes	This little fat dolphin is well known to fishermen, farmers and launch crews in the area it inhabits. Because of its quiet ways and inshore habitat almost unknown on the main coastal shipping routes, although it can be regularly observed in Lyttelton outer harbour.

A housewife living on Golden Bay (Nelson- Sth. Island) made a coloured moviefilm of her children playing with these dolphins in shallow water off the beach.

They occasionally get caught in fishermen's nets. A farmer once found one of these dolphins frantically swimming up and down near them. He discovered its mate trapped inside and freed it. After five minutes "panting", lying still in the water, it recovered and both swam on.

Mr. Fr. Robson of Taradale Napier received reports that a similar dolphin perhaps inhabits the shallow part of the Firth of Thames, a remote bay in Hauraki Gulf (Auckland).

GROUP III PLOUGH-SHARE HEADED DOLPHINS

A group of large cold water dolphins with plough-share heads, high conspicuous dorsal fins and prominent dorsal and ventral ridges along the tailstock. The colouring is black and white in a rather varied pattern. This is demonstrated in this guide by several double illustrations and a number of small sketches for the Dusky Dolphin (*Lagenorhynchus obscurus,* nr. 23).

There is some controversy among zoologists about the number of valid species. The species of the List of Marine Mammals are maintained in this guide, but the fourth has been divided into 2 for the benefit of the observer.

A sixth and tropical species *Lagenorhynchus electra* was until recently, known only from a total of 8 skulls. The stranding of a live animal in Japan (January 1963) and the subsequent description by Dr. Nakajima and Dr. Nishiwaki, made a reclassification necessary. This species is distinguished since as a new genus "*Peponocephala*" which does not belong to the Plough-share Headed Dolphins, but to the Group of the Bluntheaded Dolphins (group XIV, nr. 67).

genus	LAGENORHYNCHUS (Gray, 1846).
	WHITE BEAKED DOLPHIN, plate nrs. 19a and 19b.
species	*Lagenorhynchus albirostris* (Gray, 1846).
area	North Atlantic, north of the Gulfstream (Chart 3).
sea temperature	0–15° C.
length	9–10 feet (2.7–3.0 m).
weight	500–600 lbs; 55–65 lbs/ft (226–272 kg).
food	Fish (Whiting, Cod, Haddock, Herring) up to two feet in length.
teeth	4×25 to 26, diameter $\frac{1}{4}''$ (0.6 cm).
illustration	From description, cast and life. The beak is about $2''$ in length.
special features	The high dorsal fin is conspicuous. They are fast swimmers, who rush along the surface when breathing, making a considerable "bow wave" with head and dorsal fin; they do not jump out of the water regularly, neither has the author seen them come to a ship to play in front or alongside.
	They are decidedly larger than the next 3 species.
speed	Not recorded, but at least up to 16 and 18 knots.

breath	Not recorded.
schools	6–20; along the Norwegian coast schools of up to 1500 are reported.
biotope	Marine and oceanic.
immatures	Pairing takes place in autumn and the gestation period is 10 months. Calves are born in June and measure 4 feet (1.2 m).
notes	This dolphin regularly dies near and is washed ashore in the British Isles and North Sea coasts.
	The greater number of them are sick and old animals. They are less common in the West than in the East Atlantic. On the normal Atlantic shipping route between the Channel and America, the author recorded them in June at 50° N and 15° W, with observations of *Globicephala* and *Orcinus* in the same hour. The sea temperature was 15° C.

.WHITE SIDED DOLPHIN, plate nr. 20.

species	*Lagenorhynchus acutus* (Gray, 1828).
area	North Atlantic Ocean, north of the Gulf Stream (Chart 3).
sea temperature	0–15° C, occasionally 18° C.
length	Maximum 9 feet (2.75 m), but the majority does not seem to reach that size.
weight	Approximately 540 lbs; 60 lbs/ft (244 kg).
food	Fish.
teeth	4 × 30 to 34, small, diameter 3/16″ (0.45 cm).
illustration	From cast, photographs and life. The beak is about 2″ in length.
special features	The high, upright dorsal fin and the white (partly tinted yellow) patch on the flanks.
	They are fast swimmers, who rush along the surface when breathing, making a conspicuous bow wave with the head and dorsal fin. They do not jump out of the water regularly, and the author has not seen them come to a ship to play in front or alongside.
speed	Not recorded but estimated by the author at least 16 to 18 knots.
breath	Not recorded, but at longer intervals than most dolphins. Estimated at every minute.

schools	4–10 in the southern part of its range, up to 1000 further north.
biotope	Marine and oceanic, not necessarily deep water.
immatures	Pairing takes place in summer and autumn and calves are born from April to June. They measure 4 feet (1.2 m).
notes	On the normal North Atlantic shipping routes they can be seen in the southern part of the Labrador Current, where it meets the Gulf Stream, south of New Foundland Banks, as far west as Nantucket Lightvessel and Cape Hatteras, occasionally in the entrance of the English Channel. They are sometimes in company with *Globicephala*. They are less common in the West than in the East Atlantic.
species	PACIFIC WHITE SIDED DOLPHIN, plate nr. 21a/b. *Lagenorhynchus obliquidens* Gill, 1865, (= *L. thicolea* Gray, 1849).
area	Colder waters of the North Pacific Ocean (Chart 3).
sea temperature	0–15° C.
length	7–8 feet (2.1–2.5 m).
weight	Approximately 300 lbs; 36 lbs/ft (136 kg).
food	Fish and Squid (8–10 % of body weight daily).
teeth	Upper jaw 2 × 29, lower jaw 2 × 32.
illustration	From life and photographs. Black and white with a considerable variety in pattern. Two plates (21a and 21b) are given to demonstrate this.
special features	The black and white colouring, the high and upright dorsal fin which is also black (front) and white (rear). They approach the surface for breathing in a horizontal position and with considerable speed, making a bow wave with the head and dorsal fin. Only the fin shows and they do not jump regularly. Occasionally they come to a ship and play in the bow wave behind the vessel. Calves are very playful and jump often.
speed	Up to 22 knots.
breath	No good records; approximately every 15–20 seconds.
schools	20 to 1000.
biotope	Marine and coastal, up to about 100 miles off shore, mostly in depths of 600 to 1000 fathoms.
migration	There is undoubtedly a migration, north in the spring and

	south in the autumn.
immatures	Pairing takes place in summer and autumn and calves are born from April to August. The length is $2\frac{1}{2}$ feet (0.76 m) and the weight is 30 lbs (13.5 kg).
notes	They come further south on the eastern side of the Pacific than on the western side. The author's southernmost record is $33°30'$ N in May.

When approaching Los Angeles from the west, they can be seen during the whole year.

Far off the coast, a school of these dolphins were in close company with a school of *Lissodelphis borealis* (nr. 49) but the schools did not mix and the latter were swimming ahead. In July 1958 there were many hundreds in a cauldron of Humpback whales, Dall's Porpoises, Cormorants, Pelicans and Albatrosses; south-west of Santa Michael Island (California, USA).

All were so close together that the writer wondered how the constantly diving birds did not break their necks.

In Marineland of California, San Pedro, there were two specimens in 1958 who were taught to jump over a stick suspended about 10 feet or even more above the water.

	HOURGLASS DOLPHIN, plate nr. 22 ($= L.$ *wilsoni*, Lille 1915).
species	*Lagenorhynchus cruciger* (Quoy and Gaimard, 1824).
area	Southern Antarctic Oceans, so far not reported from the Indian Ocean sector (Chart 3).
sea temperature	$0–5°$ C.
length	5–6 feet (1.5–1.8 m).
weight	Approximately 250 lbs (113 kg).
food	No records, probably Fish.
teeth	4×28.
illustration	From cast and photographs.
special features	The very distinctive colour pattern. In the *wilsoni* the dorsal black on the tailstock does not quite reach the flukes and the black from both flanks meets between the flippers.
	They are reported to swim undulatingly like penguins. It is a very rare dolphin and a specimen of the *wilsoni* species has not yet been obtained.
speed	No record, but probably over 15 knots.

49

breath	No record.
schools	2–4 and more.
biotope	Antarctic regions, close to the pack ice.
immatures	No records.
notes	In the list of Marine Mammals this species and that mentioned in the next paragraph: the Dusky Dolphin (*L. obscurus*) are all considered one species.

Because in these dolphins the colour pattern seems to be invariably the same and the biotope of *L. cruciger* is pure antarctic, whereas *L. obscurus* is typically "temperate", they have been described in this guide in separate paragraphs. Except for a rare chance in New Zealand waters they do not occur on normal shipping routes.

DUSKY DOLPHIN, plate nr. 23a/d.

species	*Lagenorhynchus obscurus* (Gray, 1828).
	(*L. fitzroyi*, Waterhouse 1839, Sth. America), plate nr. 23b).
	(*L. australis*, Peale 1848, Sth. America), plate nr. 23c).
	(*L. superciliosus*, Lesson & Gardot 1826, Tasmania and Sth Africa).
area	Temperate southern oceans (Chart 3).
sea temperature	10–15° C, occasionally 5° C.
length	7–8 feet (2.1–2.5 m).
weight	Approximately 300 lbs (136 kg).
food	Fish and Squid.
teeth	4 × 28 to 32, diameter 1/10″ (3 mm). In upper jaw usually 2 less than in the lower jaw.
illustration	From life, cast and photographs. There is a variation in the amount of white in this species, with the result that originally several different forms were recognized as species.

Prof. Dr. E. J. Slijper brought them, in 1948, all together under *L. obscurus*.

The author describes the forms and species mentioned for practical reasons all as "Dusky Dolphins", although they might turn out to be separate species. According to Fraser for instance *L. australis* is distinctly a species.

Because in outer appearance and biotope the Hourglass Dolphin is definitely different from this dolphin, a separate paragraph (nr. 22) has been maintained in this guide.

50

All members of the Dusky Dolphin group seem to have the sweeping, curved demarcation line in common between the dorsal black and lighter flanks, from eye to anus.

In addition there are one or two white or light grey areas on the tailstock. The "light" on the flanks and undersides varies from white (in *L. fitzroyi*) to grey in other forms.

special features In the field they appear to be black. The light colours can only be noticed when they jump or swim close to the ship. The dorsal fin is high and erect, the anterior half usually black the posterior half white or grey. They are fast swimmers, coming to the surface horizontally for blowing, making a bow wave with head and dorsal fin.

When cruising along they jump out of the water and occasionally accompany ships alongside and before the bow.

speed 16 knots, maximum $17\frac{1}{2}$ knots.

breath Every 5 to 13 sec., maximum time under water about 5 minutes.

schools 5 to 30, sometimes 100 or more, usually swimming 3 to 10 abreast half a length apart. Mr. F. Robson of Taradate Napier NZ is under the impression that age groups form different schools.

biotope Off shore. Marine distribution in cooler temperate waters, occasionally close to the coast. In 1968 30 spent half an hour in the outer harbour of Timaru (South Isl. NZ) but that is exceptional.

migration Northward in autumn (April), in New Zealand to Hawkes Bay and in South Africa as far as Walvis Bay, returning southward in October-November with immatures.

immatures Born during the winter in temperate waters, late August.

The babies are smaller than $\frac{1}{3}$ of the mother.

Mating takes place in the spring at a depth of 6 feet under the surface.

notes The New Zealand and South African Dusky's are almost identical. The first have 2 prongs in the grey of the tailstock, the latter only one.

These dolphins are reported to be very friendly, at sea as well as in captivity. A mother who lost her baby because it got caught on a fishing line and drowned, remained in the area and was found dead a week later.

At the birth of a baby *Delphinus delphis* in Marineland Napier one of the Dusky's acted as midwife, whilst the father stood guard and all other dolphins swam madly round the pool. She later took over the position of "auntie" and the last few hours before the baby died of a lung infection, 4 months later, carried it on her back and flipper to the surface to breathe.

When the trainer entered their pool in skindiving equipment for the first time, the Dusky's immediately tried to bring him on their flippers to the surface.

They accept and tolerate strange dolphins of a different species in their pool without trouble or aggressiveness.

GROUP IV   # SOUTH EAST ASIA DOLPHINS

This group has been entered as separate because it deals with some of the unsolved questions in the sub-family *Delphininae*.

In 1956 Dr. F. C. Fraser described a new genus and species from a skeleton found at the entrance of the Lutong River (Sarawak, Western North Borneo).

genus   *Lagenodelphis,* Fraser 1956.

species   *Lagenodelphis hosei,* Fraser 1956.

SERAWAK DOLPHIN

Because nothing of the live animal and its habits is known it cannot be described in this book. The length of the skeleton found was 7′8″ (2.3 m) and the

teeth formula $\dfrac{44/43}{42/40}$.

The author, who navigated the Indonesian and North Borneo waters for many years, observed in that area a dolphin which cannot be traced back directly through publications or descriptions to any of the species of the Marine Mammals List. This animal is therefore very probably one unknown to science. The author recorded it and described it under the name "South China Sea Dolphin", although the name "Serawak Dolphin" would be better because the coastal waters of that country are in the centre of the region where it occurs.

This dolphin is described as nr. 24, in this guide and the author believes that perhaps already in the near future it can be proved that the Serawak Dolphin (only known from a skeleton) and the South China Sea Dolphin (only known from observations in the field) are one and the same species.

The Serawak-South China Sea Dolphin is not the only unsolved problem in this area.

In Malacca Strait lives a small dolphin, which in outward appearance and habits is rather similar to the one described above.

It is estimated to be smaller, darker and to have in proportion a longer stouter beak. The author did not find any description or record of this dolphin in the literature. He therefore described this animal as nr. 24a under the name "Malacca Dolphin", but it is possible that eventually it will be proved that both animals are the same species. Because the author is not able to indicate even the genus with certainty, he mentions in the first place the common names and their taxonomic status as *Delphininae*-species.

This is also done with another S.E. Asian dolphin. This one is small and difficult to identify. It occurs in the adjoining shallower parts of the Java Sea. Because the form and colouring are more like a *"Delphinus"* and the author believes it to be nearly related to the Red Bellied Dolphin (*Delphinus roseiventris*) of the Mollucca area (nr. 29 in this guide), it will be described with that group as nr. 29a, the "Java Sea Dolphin".

species	SOUTH CHINA SEA DOLPHIN, plate nr. 24. cf. *Delphininae species.*
area	Shallow parts of the South China Sea, north to 17° N, south to the islands of Banka and Billiton (Chart 2).
sea temperature	Above 27° C.
length	Approximately 5 feet (1.5 m).
weight	Approximately 150 lbs (78 kg).
food	Not recorded, probably Fish and Squid.
teeth	Not recorded.
illustration	From life, accurate, made from several specimens which were swimming and rolling before the bow. The grey of the back becomes gradually darker towards the flukes.
special features	They are an abundant coastal species in that area. They usually surface quietly, showing the back and dorsal fin after blowing. From a distance the back appears to be dark. The impression is that they are indifferent to ships, but when a school is near and one takes the trouble to go to the bow, if making a speed less than 13 knots, two or three are likely to be there, playing and jumping. Normally, in the school, they do not jump out of the water, and approach the ship under the surface.
speed	Maximum 13 knots, when cruising or feeding only 3 or 4 knots.
breath	Every 5 or 6 seconds.
schools	2–6, occasionally 10 and once about 20.
biotope	Coastal in depths around 15 to 20 fathoms, occasionally on the deep side of the 100 fathom line.
immatures	Calves were seen in September.
notes	This dolphin is perhaps identical with the Serawak Dolphin described from a skeleton by Fraser 1956. The Malacca Dolphin (24a) and also the Java Sea Dolphin (29a) belong perhaps to the same group. The estimated

length ($\pm$ 5 feet, 1.5 m) can be misjudged, the animals could be larger. An accurate size can only be recorded if other well-known dolphins are also present, which for this species has never been the case.

MALACCA DOLPHIN, plate nr. 24a.

species	cf. *Delphininae species.*
area	Malacca Strait between One Fathom Bank and Riou Islands (Chart 2).
sea temperature	Above 27° C.
weight	Approximately 120 lbs (54 kg).
length	Approximately 4–5 feet (1.2–1.5 m).
food	Probably Fish and Squid.
teeth	Not recorded.
illustration	From life in the field, when swimming and jumping. The author had no opportunity to study it well at close quarters before the bow.
special features	A small dolphin, with what appeared to be a stout beak. The three different colours – brown, orange and white stand out. It is quiet and difficult to detect in the usually muddy and greenish water because normally it does not jump. This species is attracted by ships and when approaching cover the last stretch to the bow with one or two leaps. Speed must be rather slow because the author never found them there when he went forward after observing the approach. Ocean going vessels normally have a speed of over 16 knots.
speed	Probably maximum about 12 knots.
breath	Not recorded.
schools	2–10.
biotope	Coastal in depths of 10–20 fathoms. The author has not observed it north of One Fathom Bank (Estuary of the Klang River).
migration	All observations from May to September and October (West Monsoon), so far none in the rest of the year.
immatures	No records.
notes	The dolphin seems to be a smaller and darker version of the South China Sea Dolphin, but jumps to reach a ship whereas the latter approaches unseen under water.

It has not been observed in the sea-arms forming the estuary of the Klang River (approaches to Port Swettenham), but it is probably the dolphin seen regularly by week-end yachtsmen when sailing off Port Dickson.

A drawback to observation is that most cargo vessels cover this stretch usually during the hours of darkness.

GROUP V   COMMON DOLPHINS

These dolphins all belong to the Genus *Delphinus* Linnaeus, 1758.
The nominate members of this genus frequent the coasts of South Western Europe and the Mediterranean in great numbers, during the larger part of the year. Because they are always attracted to ships and accompany them with a great display of leaping in the air and splashing back into the water, no seaman passing through the eastern North Atlantic, eastward through Gibraltar Strait or southward towards Dakar, can miss seeing them.
Ever since men went to sea in ships this truly was the "Common" dolphin. In the first scientific publication on whales by Linnaeus in 1758, it was, with the Common Porpoise, the only dolphin described, together with the then already better known Narwhal (Unicorn), Sperm Whale, Greenland Whale, Fin and Blue Whale.
The name "dolphin" is probably derived from the dolphin-fish (*Coryphaena hippurus*), a very colourful fish reaching 6 feet in length. There is some resemblance because it also has a steep forehead. Both animals have long been used as subjects in art, painted on pottery or sculptured as ornaments in buildings, churches, and palaces (Japan). The creatures thus portrayed are often a combination of the whale and the fish. From this common dolphin, the name became a collective one for all the smaller whales.
In shape or form there is no noticeable difference between the species of this genus and some species of the next one (*Stenella*).
The separation is osteological, one feature is a deep groove in the palate of the skull in *Delphinus*, which is absent in *Stenella*.
In the Red Bellied Dolphin (nr. 29a) a species, which is only imperfectly known, a much smaller groove is present and the last word on its taxonomical position has not yet been written.
The "Common" dolphins are fast and often leap out of the water.
The dark blue-black dorsal colour forms a conspicuous downward point to the flank underneath the dorsal fin. In this fin is a smaller or larger white area. These two characteristics make it easy to distinguish them from all other dolphins.

In the following paragraphs descriptions are given of the 5 different forms or types the author believes to be members of the *Delphinus delphis* group:

– *Delphinus delphis delphis* of the Atlantic Ocean and W. Mediterranean

57

- *D.d. ponticus* of the Black Sea
- *D. capensis* of Southern Oceans
- *D.d. subsp. (?)* of Pacific Ocean
- *D.d. subsp. (?)* of S.E. Australia
  and of 4 dolphins about which the author is not certain:
- Agulhas Dolphin S. and E. coasts of Africa
- Black-White Dolphin E. Pacific Coast
- Red Bellied Dolphin Molucca Sea and
- the Java Sea Dolphin, Java Sea.

	(ATLANTIC) COMMON DOLPHIN, plate nr. 25a.
species	*Delphinus delphis* Linnaeus, 1758.
subspecies	*D.d. delphis* Linnaeus, 1758.
area	Temperate North Atlantic Ocean, tropical Atlantic, south to Loanda and Mossamedes, occasionally to South Africa (Chart 2).
sea temperature	Above 15° C.
length	6–7 feet, maximum $8\frac{1}{2}$ feet (1.8–2.1 m max. 2.6 m). The beak is 6″.
weight	200-300 lbs, estimated 35 lbs/ft (90-136 kg).
food	Fish (Herring, Sardines, also Squid and Crayfish) 14 lbs/day.
teeth	4 × 40 to 50, very small. Diameter $\frac{1}{8}″$ (3 mm).
illustration	Life and photographs. During the winter the colours are less distinctive. The blue becomes brownish or greyish and the yellow segment paler.
	The downward point in the middle of the dorsal blue is diagnostic. In the centre of the dorsal fin there is nearly always a white triangle or circle.
special features	The most abundant species on the North Atlantic shipping routes. These dolphins always come to the bow, alongside and behind the ship to play, usually on both sides. Some remain as long as 10 to 20 minutes before the bow, mothers with calves usually remain alongside.
speed	Cruising in a favourite area 3 to 5 knots, chasing fish, accompanying or overtaking ships up to at least 25 knots.
breath	Normally 20 to 30 sec., 60 sec. swimming at full speed no exception. Longest time recorded 5 min 53 sec. under water (Marineland Napier).

schools	10 to 100 sometimes as many as 1000.
biotope	Coastal and marine, mostly 3 to 5 miles on the deep side of the 100 fath. (180 m) line.
	In mid-ocean usually in the neighbourhood of oceanic banks.
migration	*November to March:* South of 41–30° N and east of the Azores.
	March-May: Coming up from the south and entering the Mediterranean as far as Cape Bon and Italy.
	May: Partly out of the Med. again, northward along the Portuguese coast, spreading westward into the Gulfstream. Not north of latitude 50° N.
	June-August: In the Atlantic Ocean, comparatively few in the Med., reaching Cape Hatteras in America.
	They return at the end of August when there are again thousands along the Mediterranean southcoast of Spain.
	September-October: Returning to subtropical and tropical latitudes along the west coast of Africa.
immatures	Mating takes place at a depth of 6 fathoms, gestation period lasts 10 months. Calves are born over a rather extended period (June-October) and have a length of $\frac{1}{3}$rd of the mother.
notes	Schools of Common Dolphins sometimes associate with Euphrosyne Dolphins, but the schools do not mix. Many existing illustrations (including those of Norman and Fraser in their book Giant Fishes, Whales and Dolphins and of Kellogg and Slijper) seem to the author more a compilation of different species.
	The photograph of Common Dolphins in Kellogg's publication in the Nat. Geographic Magazine Jan. 1940 is excellent, but most likely taken in the Mediterranean or Atlantic Ocean.
	The author has thus far not seen this species in the Indian Ocean.
	(EAST MED.) COMMON DOLPHIN, plate nr. 25b.
subspecies	*Delphinus delphis ponticus.*

This subspecies according to the official literature, is confined to the Black Sea. This dolphin therefore must have managed to adapt itself to seawater temperatures ranging from practically arctic (5°) to tropical level (25°) although part of the population might leave the area during the winter.

Dolphins are regularly seen in the Bosporus.

A professional Dolphin hunter from the Black Sea, boatswain on a Bulgarian ship, whom the author consulted on this point agreed to general form and colour of the *Delphinus delphis* pictures shown to him, but insisted on a conspicuous black line going forward from the anus to the eye, which agrees with sketches from Russian sources.

This Dolphin, locally called the Karakash, according to the Dolphin hunter, is one of the three species hunted there (one of the others he identified as *Grampus griseus,* the third one remained unidentified and is perhaps the Greek Dolphin).

The population of the Black Sea Common Dolphin which remains in the Black Sea during the winter, accumulate so much blubber that they float when killed. From May onwards this disapppears and during the summer they are thin and light. The Dolphin hunter gave the length as 6 feet and weight 120 kg.

The author on several occasions met Common Dolphins during January in the area round Tunesia and Malta, these had the conspicuous black line mentioned above. As this species is extremely rare in the Mediterranean during the winter, the author is of the opinion that this region probably are winter quarters for at least part of the Black Sea population.

	CAPE DOLPHIN, plate nr. 25c.
species	*Delphinus capensis* Gray, 1828.
	Although the Cape Dolphin is described in the literature as a separate species. The author prefers to put it in this guide for practical reasons in the Common Dolphins Group.
area	Southern Oceans round Africa and near Australia and New Zealand (Chart 2).
sea temperature	14–20° C.
	It is difficult to indicate the differences between Common Dolphins and Cape Dolphins, especially because both "species" live in the same areas.
notes	In the Atlantic there seems to be a continuity of Common Dolphins throughout the seasons from the Gulfstream to South Africa as far as Durban in the Agulhas Current, although the author has not seen them in the cold inshore waters of South West Africa.
	In 1965: 95 dolphins, 83 of them Common Dolphins, swam

to their death in the shark nets of Durban (the other victims were Euphrosyne Dolphins (see nr. 37) and Agulhas Dolphins see nr. 27)). The identification was done through photographs sent for that purpose to the British Museum of Natural History, London. The species had not been officially recorded from that region before. The author observed, however, schools on several occasions between East London and Port Elizabeth.

The dark line from chin to flipper was very conspicuous, broad and very dark.

Excellent photographs from Common Dolphins in the SW Pacific published by the Dominion Museum Wellington and both Marinelands (Napier and Mount Maunganui) show that this population, does not have a dark line from chin to flipper or a very faint one which will be difficult to observe in the field.

The pigmentation of the tailstock is often so dark that the diagnostic V point of the dorsal blue is difficult to detect.

In both these regions the Cape Dolphin should occur. If it is a valid species or subspecies the difference in the field is not noticable. Until more definite material is published on this subject the observer has to be satisfied with an identification as Common Dolphins.

Throughout the southern Indian Ocean from Durban to halfway the Australian Bight this species has not been observed or recorded. It seems therefore unlikely that a permanent link exists between the Atlantic and Pacific populations.

Illustration 25c portrays a very dark specimen of the SW Pacific population, which might be a specimen of the Cape Dolphin. It was made 100 miles east of Hawkes Bay. The author wonders, however, whether this might be a different animal, perhaps even an offspring from a Common- and a Dusky Dolphin (nr. 23), since mating between the 2 species has been observed.

Subspecies (?) (PACIFIC) COMMON DOLPHIN. Pacific Ocean, plate nr. 26.

species   *Delphinus delphis.*

area	Three areas in the Pacific are inhabited by the Common Dolphin. NW Pacific, temperate and subtropical (Formosa, Riukiu Islands to Hawaii Islands).
	This population is probably the *Delphinus delphis bairdi* Dall, 1873 (Scheffer and Rice).
	SW Pacific, temperate and subtropical, New Zealand and Australia; SE Pacific, tropical (to temperate?) (Chart 2).
sea temperature	North above 15° C, South above 14° C.
length	6–8 feet (1.8–2.4 m).
weight	Estimated 200 to 300 lbs, 35 lbs/ft (90-136 kg) (Norris and Prescott quote 110–130 lbs, 23 lbs/ft: see notes).
food	Small Fish and Squid.
teeth	4×40 to 50, very small.
illustration	From life and photographs. The animal is almost identical to the Atlantic Common Dolphin.
	The lightcoloured segment of the tailstock is brown instead of grey and it has more black in the lower border of that area. The flippers observed in the north were almost white, in the south mixed dark and white sometimes completely dark. The dark line from chin to flipper is very faint.
special features	They are not numerous, and could only be observed at close quarters during June in latitude 32N 180E over a distance of about 100 miles.
	The SWern population is very numerous, they regularly come to play with ships, but to a lesser extent than the Atlantic Common Dolphins do.
	During March the colours appeared to be paler than in October. Of the SE population only one school was definitely seen SW of the Galapogos Islands and no special features were observed.
speed	North: 18 knots. South: over 20 knots.
breath	Every 10 to 30 sec.
schools	North: 15 to 50, 100 and sometimes more.
	SW: 10 to 200, but enormous schools have been reported off Hawkes Bay by Mr. Frank Robson of Taradale Napier covering an area of 27 miles in length and width (measured by fishermen using radiotelephony) and to the north of Bay of Plenty, similar schools by Mr. P. Howard of Marineland

	Mount Maunganui, both during summer.
biotope	Coastal, marine and oceanic, many dolphins seem to be permanently coastal. In the shallow waters of Hauraki Gulf, Bay of Plenty and Hawkes Bay they are always present.
migration	North: there is a possible migration between the Hawaiian chain and the Formosa area.

SW: only during the summer these dolphins come south to the latitude of Cook Str. and are very numerous all round North Island. In April the majority leaves for warmer regions, possibly to the northern Tasman Sea. Schools are regularly observed all the way between Australia and New Zealand.

immatures · Calves are born over an extended period in autumn and winter and have a length of $\frac{1}{3}$rd of the mother. The only baby born in captivity was in Marineland Napier 26th July 1968.

notes · They do not associate with other dolphins, but were once seen by the author in an enormous school of over 1000 animals with Euphrosyne Dolphins NE of New Zealand in September.

The "Common Dolphin" of the eastern tropical Pacific described by Norris and Prescott (1961) weighing 110–130 lbs and capable of a maximum speed of $8\frac{1}{2}$ knots must, in the author's opinion, belong to a different species. A further description of this dolphin is given under nr. 28 with the name Black White Dolphin or Baird's Dolphin.

Mr. Frank Robson of Taradale at Napier New Zealand, who was the first man to capture and train Common Dolphins, found them very intelligent and cooperative, but they display less proverbial "dolphin helpfulness" than some other species (*Lagenorhynchus* and *Tursiops*). Wounded animals are abandoned or even chased out of the school, when an immature is inadvertantly caught on a fishing line, the mother immediately deserts it.

When they had to share their pool with two Dusky dolphins they chased and attacked the newcomers continuously, slapping them on the beaks with the flukes. The Dusky's accepted this for a while, then turned round, chased the Common Dolphins and bit one in the dorsal fin. This re-established peace.

AUSTRALIAN (?) COMMON DOLPHIN, plate nr. 26a.

In December 1961 a pair of Common Dolphins accompanied the author's ship for 30 minutes through Port Philip Bay (Melbourne) playing and rolling before the bow. They were undoubtedly Common Dolphins but with a very light grey dorsal colour and they had a thin grey line from beak to anus, similar to the harness of the Euphrosyne dolphin.

The shallow inland waters where they were observed does not concur with the normal Common Dolphin-biotope. Since this observation Chief Officer K. Salwegter reported this dolphin to be very numerous around the entrance of Spencer Gulf and in the harbour of Port Lincoln, Sth. Australia.

His sketch shows a brown dorsal colour with the diagnostic downward point in the middle, the white triangle in the dorsal fin, yellow undersides and a distinct grey line from anus $\frac{3}{4}$ length along the flanks going forward.

This "Australian Dolphin" seems to be a different form, perhaps it is a separate subspecies.

AGULHAS DOLPHIN, plate nr. 27.

species	*Delphinus, Stenella* species?
area	South- and East coasts of Africa from Cape of Good Hope to Pemba Island (Chart 2).
sea temperature	15–25° C.
length	7 feet (2.1 m).
weight	$\pm$ 240 lbs (108 kg).
food	Fish.
teeth	4 $\times$ 35 to 38 (see notes).
illustration	From life and a frozen specimen of the Durban Oceanarium which had turned completely black.
	The tailstock is several shades lighter than the back.
special features	The all-brown colour and lighter tailstock is not difficult to recognize. When pursuing fish or approaching a ship they jump partly or totally clear of the surface. Some schools were afraid and fled in panic when intercepted by a ship, noticing it at a distance of about $\frac{1}{2}$ mile, in the usual manner swimming at full speed with frequent leaps in rows of 3 to

	10 and keeping the ship straight astern.
speed	About 18 knots maximum.
breath	Every 6 to 12 sec.
schools	20 to 30 and sometimes as many as 300.
biotope	Coastal outside the 100 fath. line, occasionally to depths of 50 fath.
immatures	In March and July dolphins of three-quarters length were observed but no definite immatures.
notes	During March they were seen in the Mocambique channel and between East London and Port Elisabeth. During July they were numerous around Zanzibar and Pemba Islands and a school of 300 kept pace with the ship of Cape Point. A British Pilot in Mtwara (Sth. Tanzania) noticed them almost every week when sailing outside the reefs of that harbour from March till June.

This dolphin in Durban was thought to be a *Tursiops aduncus.*

The author doubts this because *Tursiops aduncus,* the Red Sea dolphin, is restricted to the Red Sea and is greenish brown with a dark grey belly.

The lower jaw of the frozen specimen was slightly longer than the upper jaw.

The dental formula of the frozen specimen resembles that of a *Stenella* of group VIII.

The author believes for several reasons, that the Agulhas Dolphins when more is known about the animal will turn out to be a *Stenella attenuata.*

BLACK WHITE DOLPHIN or BAIRDS DOLPHIN, PACIFIC COMMON DOLPHIN (Sensu Norris and Prescott), plate nr. 28.

species	*Delphinus bairdi* Dall.
area	Eastern Pacific coast from Costa Rica to Gulf of California and probably further north to Los Angeles (Chart 2).
sea temperature	27–31° C.
length	5–6 feet (1.5–1.8 m).
weight	110–113 lbs (50–59 kg). See notes.
food	Small Fish (Anchovies, Sardines, etc.) and probably Squid.
teeth	Not recorded.
illustration	Made from life and photographs of large schools in the pub-

lication: Results of the "Puritan American Museum of Natural History Expedition" to Western Mexico, the Gulf of California in 1960.
This dolphin is described by Norris and Prescott as the "Pacific Common Dolphin".
The white area in the dorsal fin characteristic for the *Delphinus delphis* is present. The black V under the dorsal fin is absent.

special features	The sharply contrasting black and white colours, the white area in the dorsal fin. When the author met this dolphin it made no effort to come towards the ship and play. Cruising, they often jump out of the water.
speed	Cruising 5–6 knots; maximum $8\frac{1}{2}$ knots (N & P).
breath	Not recorded.
schools	10–20 and more.
biotope	Coastal and marine.
immatures	Pairing in spring, calves are born in the winter.
notes	The author met and recorded this dolphin several times in the neighbourhood of the extensive banks approximately 100 miles south-west of the Gulf of Nicoya. Not recognizing a known species he described them in his notes as an unknown species.

Later he recognized this dolphin in the photographs of the Puritan Expedition. In his opinion this dolphin is not a "Common Dolphin".
Comparing figures 25 and 26, the author is convinced that the Atlantic Common Dolphin and the North Pacific Common Dolphin both are 'nearly related" Common Dolphins. These Common Dolphins, however, are bigger (6–7 feet), heavier – estimated at over 200 lbs – and much faster swimmers (over 18 knots), than the Pacific Common Dolphin – the *Delphinus bairdi* of Norris and Prescott: 5–6 feet, 110–130 lbs, maximum speed $8\frac{1}{2}$ knots. The author therefore used as Common name for this animal not Pacific Common Dolphin, but Black White Dolphin.
Because of these differences and the questions concerning its taxonomy, this dolphin has a paragraph of its own.

RED BELLIED DOLPHIN, plate nr. 29.

species	*Delphinus* cf *roseiventris* Wagner, 1846.
area	Molucca's to Torres Strait (Chart 2).
sea temperature	28–31° C.
length	4 feet (1.2 m).
weight	Estimated at 110 lbs (50 kg).
food	Probably Fish and Squid.
teeth	Upper jaw = 2 × 48; lower jaw = 2 × 45.
illustration	This species is only known from 3 skulls and a vague description. The illustration in this guide is of a small dolphin observed in the area mentioned, including the seas around Celebes Island. This little dolphin, however, has a white to yellow belly, not pink or red as might be expected. The dolphin of plate nr. 29 therefore is not certain to be a *D. roseiventris*.
special features	Small size. They leap above the surface, but are not specially attracted to ships. The author has seen them come to the bow, but they only stayed a very short time. When they leap, the belly can often be seen clearly.
speed	Probably maximum 8–10 knots.
breath	Not recorded.
schools	6–50.
biotope	Coastal in deep water, near coral reefs?
immatures	No records.
notes	There is a shallow groove in the palate of the skull in *D. roseiventris* which places this species between *Delphinus* and *Stenella*.

A red or pink belly is very common for dolphins in the tropics, many species have it, including some *Tursiops* and *Stenella* sp. It is therefore not a valuable diagnostic point.

This dolphin of the Molucca area seems to be very similar to the Java Sea Dolphin described in the next paragraph.

The biotope in the shallow muddy Java Sea is, however, very different from that in the Molucca's. It must be borne in mind that the Torres Strait is also shallow.

For future research it is important to compare the data of the Red Bellied Dolphin (29) and the Java Sea Dolphin (29a), with those of other SE Asian Dolphins, especially the South China Sea Dolphin (24), and Malacca Dolphin (24a).

JAVA SEA DOLPHIN, plate nr. 29a.

species	*Delphininae spec.?*
area	Java Sea from the North coast of Java to Banka & Billiton, mainly coastwise between Semarang and Tjeribon (Chart 2).
sea temperature	28–31° C.
length	Approximately 4 feet (1.2 m).
weight	Approximately 100 lbs ($\pm$ 50 kg).
food	Probably Fish and Squid.
illustration	From life, before the bow in Semarang roads.
special features	It accompanies small boats on their way into the shallow harbours and occasionally ships leaving the roads. They approach unseen under water, but jump in the bow wave, and remain there for up to ten minutes.
speed	Maximum 12 knots.
breath	Not recorded.
schools	2–10.
biotope	Shallow coastal waters.
migration	They were mostly seen from January to May (West Monsoon), only occasionally between June and December.
immatures	In September, in a school of 20, 2 immatures were observed.
notes	It is well known to the Javanese fishermen, who call it Ikanbabi (fishpig), and do not catch it because they are afraid of the "many teeth".
	The Java Sea Dolphin seems to be very similar to the Red Bellied Dolphin (29), for that reason it is described with the genus *Delphinus*.
	Earlier a comparison is made with the Malacca Dolphin (nr. 24a).

GROUP VI, VII, VIII AND IX: STENELLA-DOLPHINS

These dolphins belong to the genus Stenella [= *Prodelphinus,* (Van Beneden & Gervais, 1877), Gray, 1866.]
Genus *Delphinus* is separate because these dolphins have a deep groove in the palate which other dolphins do not have.
In the genus *Steno,* the dolphins have teeth with rough instead of smooth crowns. Practically all other Ocean Dolphins are brought together in the genus *Stenella,* because too little is known about the species to make a better division.

As many of them are rare, often live in remote ocean areas and none have commercial value, it will take quite a long time yet to collect sufficient evidence and material for science to find out which species and subspecies exist and where they live.
From what the author has seen and recorded, he describes in this chapter how the species fit or do not fit into the rather accurate and practical sub-division suggested by Scheffer and Rice in their Marine Mammels List.

GROUP VI – LONGBEAKED DOLPHINS, with a long to very long slender beak, a rather uniform brownish coloration on back and flanks, white or light coloured underneath.
In this group are paragraphs for:

– *Stenella longirostris*	Carribbean, Atlantic, Ind. Ocean and Centr. Pac. Ocean
– *Stenella longirostris subspec.?*	Arabian coast and Persian Gulf
– *Stenella longirostris subspec.?*	Galapagos Islands
– *Stenella alope*	Coasts of Ceylon and Bay of Bengal
– *Stenella microps*	North Pacific Ocean, Hawaii and the west coast of Mexico.

They seem to be more coastal and marine than oceanic, and mostly in deep water.

GROUP VII – HARNESSED DOLPHINS, with a conspicuous black or very dark line along the flank from eye to anus, comparable to a horse's harness, normal beaks. They have a uniform coloration of dark blue (?) or brown dorsally, lighter on the flanks and white to pink underneath. In this group are paragraphs for:

– *Stenella caeruleoalba*	South Atlantic and North Pacific
– *Stenella euphrosyne*	North Atlantic and South Pacific
– *Stenella euphrosyne*	North Pacific
– *Stenella euphrosyne*	Tropical oceans.

These dolphins are mainly oceanic.

GROUP VIII – SPOTTED DOLPHINS. Dark coloured dolphins, with shorter and stouter snouts, the body is covered all over with small white or light spots. In this group are paragraphs for:

– *Stenella plagiodon*	Atlantic Ocean, tropical Gulfstream
– *Stenella graffmani*	Pacific Ocean – Panama to Mexico
– *Stenella frontalis*	Tropical Atlantic (and Indian?) Ocean
– *Stenella attenuata*	Tropical Atlantic and Pacific Ocean.

These seem to be mainly coastal with even a possible preference for life inside the 100 fathom line.

GROUP IX – NARROW-SNOUTED DOLPHINS. Small dolphins with long, narrow (sideways compressed) beaks, rather similar to the last two species of group VIII but mainly black or dark grey with some variety in pattern and no spots.

In this group are 2 paragraphs for:

| – *Stenella malayana* | from S.E. Asian waters |
| – FLORES SEA DOLPHIN | a provisional name to indicate the area where it was observed. |

These dolphins are coastal and nearly always in deep water.

GROUP VI	LONG BEAKED DOLPHINS
	LONG BEAKED DOLPHIN, plate nrs. 30 and nr. 31.
species	*Stenella longirostris* (Gray, 1828).
area	Tropical oceans (Indian, Atlantic-Carribbean, Pacific) (Chart 3).
sea temperature	24–30° C.
length	Approximately 6 feet (1.8 m).
weight	Approximately 150–200 lbs (78–90 kg).
food	Probably Fish and Squid.
teeth	4 × 50.
illustration	From life in the field.

The general colour is dark brown, lighter on the flanks to

white on the underside (in august pure pink). Those in the Atlantic Carribbean and Indian Oceans had a separate, not much different shade on the flanks; in the Central Pacific there were only two different colours.

special features   The long beak and the brown colour. They are afraid and always swim away when a ship comes near them.
In those cases the ship is kept right astern (which makes observation difficult) especially as when escaping they prefer to reach a position between the ship and the sun.

speed   Estimated 16 or 18 knots, probably faster if necessary.

breath   No record.

schools   10–50 and 100, 20–30 in the Central Pacific.

biotope   Mostly marine and oceanic, occasionally coastal.

immatures   Pairing probably throughout the year. In the Carribbean there were calves $\frac{2}{3}$ length in April.
In the Indian Ocean calves of $\frac{1}{2}$ and $\frac{2}{3}$ length in January. Off the Marqueses Islands calves of less than $\frac{1}{2}$ length to $\frac{2}{3}$ in June, and at the Galapagos Islands, newborn calves in August.

notes   Figure 30 was made in the Indian Ocean and Carribbean Sea. Plate 31 in the Marqueses Islands, Pac. Ocean).
They are especially abundant in the Central Pacific Ocean between the Tuamotu Islands and 100° W, which is the South Equatorial Current.
Schools of these dolphins sometimes enter the large bays of Pacific Islands and the inhabitants of Nukuhiva (northern Marqueses) discovered that by cutting them off from the sea with their canoes and then frightening them with the noise of high pitched rattles, at least 10 to 12 animals of the school became so panicky that they swam straight on the beach, where all efforts to refloat them again always failed. Because this was just a sport and the dolphins were not eaten or used otherwise, the French Government has now forbidden the game. (Information Monsieur Ch. Coulon, Tahiti, who spent 10 years trading with a schooner in the Central Pacific area).
In the Marine Mammals List S. *microps* and S. *alope* are considered conspecific with S. *longirostris*.
Because in the field they appear to have a different colour

pattern and habitat, both are described in a separate paragraph.

SPINNER DOLPHIN, plate nr. 32.

species	*Stenella microps* (Gray, 1846).
area	Tropical North Pacific (Chart 3).
sea temperature	22–30° C.
length	Approximately 7 feet (2.1 m).
weight	Approximately 200 lbs (90 kg).
food	Fish (15 lbs a day).
teeth	4×50.
illustration	From life and photographs. The belly is whiter than the illustration indicates, but only on the flat part. The chin is pinkish.
special features	The exceptionally long thin beak. When cruising along the surface in a school they jump reguarly and above the water, spin round very rapidly before falling back often horizontally or even tail first, with a considerable splash. In the author's experience they are indifferent to big ships, but are known to play before and follow fishing vessels.
speed	Approximately 15 knots.
breath	Not recorded.
schools	20–100, and more.
biotope	Deep water, coastal and marine. They are best known from the Hawaiian Islands and Tres Maria's (Gulf of California). The author also recorded them near the Revilla Gigedo Islands and off the coast of Costa Rica.
immatures	No records.
notes	There are several captive animals in the "Sea Life Park" – Oahu, Hawaiian Islands, where beautiful coloured photographs were taken and published.

Also a coloured film was taken of a large school. In the Marine Mammals List this dolphin is treated as conspecific with the *S. longirostris*.

Because they appear to be different and the author is able to distinguish between both dolphins without difficulty he mentioned the Spinner Dolphin as a species and has given it a separate paragraph.

BENGAL or CEYLON DOLPHIN, plate nr. 33.

species	*Stenella alope* Gray, 1850, (= *Stenella longirostris alope*).
area	Indian Ocean around the south point of India, Ceylon and perhaps the Andaman Sea (Chart 3).
sea temperature	26–30° C.
length	Approximately 6 feet (1.8 m).
weight	Approximately 200 lbs (90 kg).
food	Fish and probably Squid.
teeth	4×50.
illustration	From life and photographs. At sea the dolphin appears to be all dark. The yellow is mostly on the flat of the belly and can only be seen when it swims and jumps sideways.
special features	The long snout and dark streak from eye to flipper. Often come to the bow and alongside ships and play the same way as the Common Dolphin. As many as 10 can keep station together in front of a ship's bow and "spin" when jumping.
speed	Over 20 knots.
breath	Probably every 10–30 seconds.
schools	10–50.
biotope	Coastal, sometimes marine. Inside and around the 100 fathom line round Ceylon, probably also in the Nicobar Islands and around the north point of Sumatra.
immatures	Newborn calves, probably of this species were observed in April near Nth. Sumatra.
notes	In the last mentioned area also live *S. malayana* (nr. 45) and *Tursiops* (nr. 60). The schools do not mix, but unless close to the ship it is difficult to distinguish one species from another. This dolphin is occasionally for sale on the Ceylon fish-markets. The Bengal Dolphin is sometimes looked at as conspecific with *S. longirostris*. Because they appear to be quite different, the author has given *S. alope* a separate paragraph.

ARABIAN DOLPHIN, plate nr. 34.

species	*Stenella longirostris subspec.?*
area	Arabian coasts from Aden round the peninsula to Kuwait, perhaps also down the East African coast (Chart 3).
sea temperature	20–30° C.

length	Approximately 6 feet (1.8 m).
weight	Approximately 250 lbs (113 kg).
food	Probably Fish.
teeth	About 4×50.
illustration	From life, made and perfected during successive voyages through the area.
special features	Very playful and quite often come to the bow and alongside to play for 10 to 20 minutes.
	Under these circumstances their behaviour is very much like *Delphinus delphis* and *Tursiops,* making especially spectacular leaps and jumps in the spreading bow wave alongside and behind the ship.
	Like all dolphins which use ships, they ride the crest of these waves, then make a somersault before starting on the next "run". They do not "spin".
speed	Up to 20 knots.
breath	Every 10 to 20 seconds.
schools	10–20, sometimes many hundreds.
biotope	Coastal and marine, along and inside the 100 fathom line.
immatures	Newborn and half-grown calves were seen in February and June.
notes	This playful dolphin will always be seen on any voyage to the Persian Gulf, concentrations of thousands have been observed around Ras Fartak and on either side of Strait Hormez.
	This dolphin undoubtedly belongs to the group of *Stenella longirostris,* but has distinctly its own colour pattern and area.

GALAPAGOS DOLPHIN, plate nr. 35.

species	*Stenella longirostris subspec.?*
area	The colder waters of the Humboldt current, near and westward of the Galapagos Islands (Chart 3).
sea temperature	20–23° C, occasionally higher.
length	6–7 feet (1.8–2.14 m).
weight	$\pm$ 250 lbs ($\pm$ 110 kg).
food	Probably Fish.
teeth	Not recorded.
illustration	From life.

special features	They are afraid of the ship and difficult to approach, only jump when intercepted and trying to escape.
	The sweeping demarcation line between the dark brown dorsal colour and lighter flank is easy to recognize.
	From far off it makes them resemble Dusky Dolphins.
	The colouring is very similar to the Arabian Dolphin nr. 34.
speed	To about 18 knots maximum.
breath	Every 10 to 20 sec.
schools	10 to 200, several smaller schools often within 1 or 2 miles of each other, the animals swim close together and are strung out in a long front giving the school a depth which is only a fraction of the width.
biotope	Marine round the Galapagos Islands, rare to the NE of this group and not observed between the islands, where a *Tursiops* with a very short beak occurs in great numbers.
immatures	Not recorded.
notes	This dolphin has a very long beak and probably is a subspecies of *S. longirostris*. The same area is inhabited by the Euphrosyne Dolphin (nr. 37), who is less shy but also swims away from ships, and the *Tursiops nuuanu* (nr. 62), smaller and completely brown. Euphrosyne schools are roughly oval in shape and the animals maintain a distance of one length between them. *Nuuanu* schools are also oval in shape, but extremely compact, the dolphins almost touching each other which gives the sea in an escaping group the appearance of boiling.
	The schools of the Galapagos Dolphins therefore can easily be distinguished.

GROUP VII	HARNESSED DOLPHINS
species	BLUEWHITE DOLPHIN, plate nr. 36.
	(a) *Stenella caeruleoalba* (Meyen, 1833) and
	(b) = *S.c. styx* Gray, 1846.
area	(a) Western South Atlantic (Chart 4).
	(b) South African waters.
	(c) Also reported from the NW Pacific Ocean by Prof. Nishiwaki of Japan.
sea temperature	5–15° C.

length	Probably about 6 feet, the type specimen was only 3′10″ (1.8 m; 1.17 m).
weight	Approximately 150 lbs, estimated 35 lbs/ft (68 kg).
food	Probably Fish.
teeth	4×50.
illustration	From description and cast:
	(a) was recorded as dark steel blue with white belly and snout;
	(b) is black where (a) is blue. In this dolphin the narrow dark line from eye to anus does not run the full length but stops in front of the anus and near the end forms a little fork.
special features	The narrow dark line, which can only be seen when the animal jumps out of the water.
speed	Not recorded.
breath	Not recorded.
schools	Not recorded, probably 10–30.
biotope	(a) Falkland Current, marine and coastal;
	(b) Benguella Current (?)
immatures	No records.
notes	Information on this species is extremely scarce.
	If the type specimen of (a) and the specimen of (b) on which the description was based, was dead for any length of time, it is quite possible the original life colour was different. Dolphins lose their coloration very quickly after death and turn dark blue and black (see also nrs. 27, 38).

	EUPHROSYNE DOLPHIN (North Atlantic form.), plate nr. 37.
species	*Stenella euphrosyne* (Gray, 1846) (also *S. caeruleoalba euphrosyne*).
area	Two separate areas: Temperate North Atlantic Ocean and Mediterranean Sea and in temperate SW Sth. Pacific Ocean (Chart 4).
sea temperature	5–18° C.
length	6–7 feet (1.8–2.1 m).
weight	200–300 lbs; 36 lbs/ft (90–136 kg).
food	Fish.
teeth	4×45.
illustration	From life. The dorsal brown is sharply separated from the

white of the flanks. The white streak invading the dorsal black is only conspicuous when the animal is close.

special features	The "harness-line" and the white streak towards the dorsal fin. They are not afraid of ships, rarely come to the bow but often play astern.
speed	Over 20 knots. The author witnessed a school overtaking his ship which was doing 21 knots at that time.
breath	Every 10 to 20 seconds.
schools	Roughly oval shaped. From 2 to 50 in the South Pacific, sometimes more, the animals well spaced.
biotope	Oceanic and marine. North of 35° N in the North Atlantic, and occasionally in the Western Mediterranean. Oceanic and marine North and North East of New Zealand.
immatures	Newborn calves seen at Azores in May, Gibraltar in July; immatures of $\frac{3}{4}$ length were seen in August, December and April.
notes	In the North Atlantic, they associate with the Common Dolphin but the schools do not mix. When the Common Dolphin veer towards a ship to play, the Euphrosyne sometimes follow suit.
	In the South Pacific they were observed once together in a school of about 1000 dolphins with *Delphinus delphis*, probably all chasing the same fish.

EUPHROSYNE DOLPHIN (North Pacific) form., plate nr. 38.

species	*Stenella euphrosyne* (Gray, 1846) (also *S. caeruleoalba euphrosyne*).
area	Temperate North Pacific Ocean (Chart 4).
sea temperature	5–15, perhaps 25° C.
length	6–7 feet (1.8–2.1 m).
weight	200–300 lbs; 36 lbs/ft (90–136 kg).
food	Fish.
teeth	4 × 45.
illustration	From life and photographs. The coloration on the flanks is more varied than in the North Atlantic form; the harness-line is heavier and widens considerably around the vent.
	The white underside is orange-pink and the white streak from the flank into the dorsal black not very conspicuous.
special features	They are afraid of ships and try to swim away on their

original course. When overtaken they veer off at right angles to the ship. When fleeing they jump continuously and with increasing speed return to the water with steeper angles, which gives a good opportunity to see the coloured belly and the black line going around it.

speed	Approximately 17 knots maximum.
breath	Probably every 10 to 20 seconds.
schools	Roughly oval shaped of 10–40 animals well spaced.
biotope	Oceanic and marine.
immatures	Not recorded.
notes	Norris and Prescott give a description of this dolphin in "Pacific Cetaceans". They received the animal a few hours after it had beached itself near Los Angeles and was shot by the policeman who found it. It was then a very dark brown, the policeman reported it as brown and by the time it was frozen it was dark steel blue. If the same happened to the type-specimen of the Bluewhite Dolphin, it might well turn out yet that it is also brown when alive.

This dolphin was seen regularly by the author on a Pacific crossing from the Milwaukee Banks (extention of Hawaiian chain) to Los Angeles. There is no sharp demarcation between the area of this dolphin (light bellied) and the next (dark bellied). Light bellied schools were occasionally seen as far south as the Carolines and west of the Philippine Islands. The dolphin is undoubtedly the same, but the tropical version is darker.

EUPHROSYNE DOLPHIN – tropical form, plate nr. 39.

species	*Stenella euphrosyne* (Gray, 1846).
area	Tropical oceans (Chart 4).
sea temperature	20–30° C.
length	6–7 feet (1.8–2.1 m).
weight	200–300 lbs (90–136 kg).
food	Fish and probably Squid.
teeth	4×45.
illustration	From life. Back is dark brown, flanks various brownish and whitish stripes above the harness-line, undersides dark pink.
special features	From a distance this dolphin looks all dark, especially as in the tropics the sun is usually very high and the lighter under-

side in the shadow. They are scared and only on rare occasions come to play alongside a ship.
In the author's experience only once in ten years.
When swimming at maximum speed they frequently jump, emerging from the water at an angle of about 30° and returning at an angle of about 45°. In big schools when playing or perhaps pairing, well away from ships, these dolphins often leap to six feet and more out of the water in any direction, falling back flat with considerable splashing.

speed	Up to 15½ knots.
breath	Every 5 seconds, probably capable of staying under water longer.
schools	Roughly oval shaped 6–100 animals well spaced.
biotope	Marine and sometimes oceanic. Seems to be more attracted to island groups (Maladives, Carolines and Marshall Islands, etc.), than to open ocean.
immatures	Pairing probably all the year round. Newborn calves were observed in February, May and November. There seems to be a peak in immatures during the northern summer.
notes	In the field it is almost impossible to distinguish them from other brown dolphins in the same area such as *S. longirostris* and *Tursiops nuuanu*. As mentioned in the preceding paragraph there is probably continuity from temperate to tropical oceans, and very likely also between the Pacific and Indian oceans.

GREEK DOLPHIN, plate nr. 40.

species	*Stenella spec.?*

This dolphin was observed and recorded by the author in the Eastern Mediterranean, and in outward form and appearance cannot be identified with any of the known species.
There is an unmistakable similarity between this animal and the Euphrosyne Dolphin, but the system of black lines is completely absent. The animal is smaller, more chubby, and has a shorter, stouter snout.

area	Eastern Mediterranean, north of 34° N, and east of Sardinia. Perhaps further west (Chart 4).
sea temperature	15–18° C, perhaps also higher.

length	5–6 feet (1.5–1.8 m).
weight	Estimated at 150–200 lbs (68-90 kg).
food	Probably Fish and perhaps Squid.
teeth	No records.
illustration	From life. The dorsal brown is well defined to behind the dorsal fin. The white streak is similar to that in the Euphrosyne dolphin. The rest of the body is dirty white, discoloured by smoky brownish and grey smudges behind the dorsal fin. The underside is white.
special features	A short plump dolphin, with a rather short and stout snout. It rarely pays attention to ships. When surfacing it shows a good deal of the body, but does not often jump completely out of the water.
speed	Up to 15 knots.
breath	Not recorded.
biotope	Coastal and occasionally marine – in deep water.
schools	2–10, occasionally 50.
immatures	Calves of $\frac{1}{3}$ to $\frac{1}{2}$ length were observed in January near the Stromboli. Possibly newborn calves in the same area in October, but these might have been Euphrosyne Dolphins.
notes	The shape of this dolphin is practically the same as that of the "Senegal Dolphin" described as *Stenella spec.* (?) nr. 47.

GROUP VIII SPOTTED DOLPHINS

GULFSTREAM SPOTTED DOLPHIN, plate nr. 41.

species	*Stenella plagiodon* (Cope, 1866).
area	Gulfstream, Carribian to Cape Hatteras (Chart 4).
sea temperature	25–31° C.
length	7 feet (2.1 m).
weight	280 lbs; 40 lbs/ft (127 kg).
food	Fish.
teeth	4×34 to 37, pointed but square at the level of the gum, diameter $\pm$ 5 mm.
illustration	From life and cast. The white spots can only be observed when the dolphin is very close to the ship. The pattern is much finer and more intricate than painted here. The front part of the snout is white.
special features	The outward form of this dolphin is more like a *Tursiops*

than a *Stenella*. The dorsal fin is high in some larger specimens. It often comes to the bow of ships and plays alongside. On these occasions it jumps frequently, when away from ships only rarely.

speed	Over 18 knots.
breath	Every 8–12 seconds.
schools	2–10.
biotope	Coastal and marine in the area, inside the 100 fathom line.
immatures	Young calves were observed in June (half length).
notes	Dr. R. Kellogg used the name "longsnouted dolphin" for this species in his article in the National Geographic Magazine in January, 1940.

Because this dolphin has no "conspicuous" long snout in this guide the preference has been given to the common name "Gulfstream Spotted Dolphin".

The author has observed it several times between the island of Curaçao and the Panama Canal inside the 100 fathom line near Monjes Islands off Maracaibo Bay, and between Charleston and Cape Hatteras, and it was seen continously during a passage across Campeche bank.

He has the impression that it is more common on than outside the continental shelf.

GULF OF PANAMA SPOTTED DOLPHIN, plate nr. 42.

species	*Stenella graffmani* Lönberg 1934.
area	Tropical East Pacific Coast from Acapulco (Mexico) to Gulf of Panama (Chart 4).
sea temperature	26–28° C.
length	7 feet (2.1 m).
weight	Approximately 280 lbs; 40 lbs/ft (127 kg).
food	Fish.
teeth	4 × 40 to 43. Usually more in the upper than the lower jaw. Diam. ± 2.5 × 4.8 mm.
illustration	From life. The animal appears to be dark brown, slightly lighter on the flanks and underside. Close by the general colour is definitely purplish with big round lighter patches on the flanks. The spots are much finer and form a more intricate pattern than painted here.

special features	Rather similar to a *Tursiops* but with a much lower dorsal fin. It regularly comes to the bow and alongside of ships to play. The white spots can only be seen within a distance of about 100 yards. It jumps out of the water to reach a ship and whilst playing with it. The dorsal fin is comparatively small.
speed	Approximately 14 knots.
breath	Every 10 seconds to one minute.
schools	6–50, and sometimes up to 100.
biotope	Along and inside the 100 fathom line in the area. The continental shelf is rather narrow along that coast. Regularly seen amongst the ships anchored in Balboa roads.
immatures	Not recorded.
notes	This dolphin nearly always meets the ships coming in to Panama Bay from the Pacific Ocean. The author for several years thought he observed the *Tursiops gilli* (nr. 61), but wondered about the spots until on one voyage he had a school around the bow and alongside for nearly 20 minutes and could then recognize the species. Near Acapulco they associate with *T. gilli* but the schools do not mix.

	ATLANTIC SPOTTED (= BRIDLED) DOLPHIN, plate nr. 43.
species	*Stenella frontalis* (G. Cuvier, 1829). [= *S. dubia* (G. Cuvier, 1812)]. [= *S. fraenata* (G. Cuvier, 1836)].
area	Tropical Atlantic Ocean (and Indian Ocean?) (Chart 4).
sea temperature	26–30° C.
length	6–7 feet (1.8–2.1 m).
weight	Approximately 250 lbs (113 kg).
food	Fish.
teeth	4 × 37 to 38 diam. ± 3 mm.
illustration	From life of what the author considered to be this species. The exposed part of the back, which regularly comes above the surface during normal swimming and breathing is darker then the other parts of the body. Nearly all tropical dolphins have this. The painted spots are only indicative.
special features	The dorsal fin is shaped like that of a *Tursiops truncatus*, but the snout is longer and more slender.

This dolphin in form and pattern is rather similar to the Red Sea Dolphin (*Tursiops aduncus*) (63).

It comes to the bow to play. The height of the beak is greater than the width.

speed	Over 18 knots.
schools	20–50.
breath	Not recorded, about every 6–10 seconds.
biotope	Deep water.
immatures	No records.
notes	The only occasion the author could observe this dolphin clearly was between the islands of Aruba and Curaçao of the Neth. Antilles.

Most specimens known to science were obtained from the eastern tropical Atlantic (Cape Verde Islands). The given name Bridled Dolphin is not a very good one because the bridle is at best very faint and absent in most animals.

The author would prefer the name Atlantic Spotted Dolphin. He has never seen it in the Indian Ocean, although 2 specimens were officially recorded from that area.

Measurements of this species and the next (*S. attenuata*) border on each other, *S. frontalis* is smaller and has a shorter, broader beak.

NARROW SNOUTED DOLPHIN

species   *Stenella attenuata* (Gray, 1846).

Another spotted dolphin, closely resembling the *S. frontalis,* the spots seem to be restricted to the flanks only and perhaps they are not a distinguishing feature for the species.

The only dolphin which answers the description of this species was seen by the author near the Philippine Islands. The particulars of that animal are given in this paragraph. The author should like to use the name "Philippine Dolphin" because he is not sure that his Philippine Dolphin is identical with the Narrow Snouted Dolphin described from other regions, where he never observed one.

PHILIPPINE DOLPHIN, plate nr. 44.

species	*Stenella attenuata?*
area	Tropical Western Pacific, Philippine Islands to the North coast of Taiwan (*S. attenuata* also: trop. East Atlantic and central Pacific Ocean) (Chart 3).

sea temperature	26–31° C.
length	6–8 feet (1.8–2.3 m), slightly longer than *S. frontalis*.
weight	Approximately 280 lbs (115 kg).
food	Fish and probably Squid.
teeth	4 × 35 to 44, diam. 2 à 2,5 mm.
illustration	From life. Only the sides are spotted, roughly illustrated in the plate. Round the anus is an almost circular pink patch.
special features	The snout is comparatively long and sideways compressed. The schools cruise along very quietly and usually will only be noticed when they take action to flee from an approaching ship, jumping along the surface. In these cases they "close ranks" and keep the ship (noise) straight astern, making a 180 degree turn by the time the ship has passed. The dorsal fins have a great variety in shape from the normal delphinus type to a narrow sharp sickle or a flat triangle. Usually some come to the bow, but approach under water to stay for only a few seconds without breathing. They veer off to a safe distance to surface, then come back again but never for very long.
speed	Approximately 17 knots.
breath	Every 6–10 seconds.
schools	20–30.
biotope	Deep coastal waters – the Philippine Islands are a perfect type locality.
immatures	Not recorded.
notes	It is regularly seen on all the main shipping routes through the Philippine Islands between Manila, Iloilo and Sebu.

GROUP IX	NARROW SNOUTED DOLPHINS
	MALAY DOLPHIN, plate nr. 45.
species	*Stenella malayana* Lesson, 1826.
area	From Northern Sumatra, round outside of Indonesia to Solomons Islands and Paracel Islands (Chart 3).
sea temperature	27–31° C.
length	Approximately 7 feet (2.1 m).
weight	Approximately 280 lbs (127 kg).
food	Fish and probably Squid.
teeth	4 × 39.

illustration	From life of what the author considers must be this dolphin. Uniformly dark grey, sometimes a narrow white line is reported along the belly. Immatures are lighter grey with white undersides.
special features	The snout is long and narrow as in the *S. attenuata*. They do not jump when normally cruising along, but do when fleeing from a ship and on the few occasions when they come to the bow to play.
speed	Approximately 15 knots.
breath	Every 7–10 seconds.
schools	10, 30 and 50, occasionally more.
biotope	Coastal and marine deep water but they also come on the continental shelf to depths of 20 fathoms. The north coast of New Guinea is a good type locality. They are specially numerous round islands.
immatures	Probably throughout the year; newborn calves were seen in January and June.
notes	In the mentioned area the author has most frequently observed them between the Philippines and the east side of New Guinea. In northern Malacca Strait they are difficult to distinguish from the other dark coloured dolphins (*S. alope*? and *Tursiops*). Because it can be expected that there is continuity, the connection between the extreme limits very likely goes through the Moluccas, then south of Java and east of Sumatra. They have not been observed by the author however in the shallow southern part of the Malacca Strait, Java Sea and South China Sea.

Near the entrance to the port of Manokwari on the north coast of New Guinea, the author went through a school of about 50 of these dolphins milling round slowly together with *Grampus griseus,* all lobtailing and making vertical dives on the 100 fathom line which was less than 1000 yards off the coast.

In the Marine Mammals List they are considered conspecific with the *S. attenuata*.

Because they lack the spotted flanks which is stated to be characteristic for that species, they have been given a separate paragraph in the guide.

FLORES SEA DOLPHIN, plate nr. 46.

species   *Stenella malayana subspec.?*

This dolphin has the same body-form and the same narrow snout as the Philippine Dolphin (44) and the Malay Dolphin (45). The colour pattern is very different except for the round pink patch near the anus. Because it seems likely these three dolphins are related to each other the paragraph is entered here.

area	Eastern Java Sea and probably Flores Sea (Chart 3).
sea temperature	27–30° C.
length	Approximately 6 feet (1.8 m).
weight	Approximately 250 lbs (113 kg).
food	Probably Fish and Squid.
teeth	No records.
illustration	From life, South-east of Madura Island, where a small school accompanied the ship for about 15 minutes.
special features	The black dorsal colour, which shows a definite point underneath the dorsal fin, and the long thin snout. This dolphin occasionally comes to the bow, and swims in front of the ship, but like the Philippine Dolphin does not surface there for breathing.
speed	15 knots.
breath	Every 7 seconds.
schools	Up to 10 animals, the formation is spread out.
biotope	Coastal, deeper water. The least depth in Madura Strait was 30 fathoms.
immatures	In September in a school there was at least one newborn calf.
notes	The author observed these dolphins in the above mentioned Madura Strait and off the reefs of South-west Celebes. He believes this is one of the species which inhabit the waters round the lesser Sunda Islands in (combined) schools of thousands.
	These impressive numbers are recorded from Str. Alor and also round Buru and Ceram Islands by Mr. K. Salwegter on several voyages.
	The baby mentioned above tried to take station before the bow and frantically waved right and left about three or four times before it had to give up for lack of speed.

SUBFAMILY DELPHININAE

UNIDENTIFIED DOLPHINS

Originally there were 4 dolphins in this group. They were brought together because the author has seen them on several occasions and well enough to make an illustration, but they could not be placed in, or added to one of the known genera.
One or more specimens have to be caught, studied and identified by a marine mammologist, before a decision can be taken.

The Greek Dolphin described as nr. 40 *Stenella spec.*, looking very much like a *S. euphrosyne* without a harness has been placed with that genus.
The Black White Dolphin is almost certainly the *Delphinus bairdi* from the tropical eastern Pacific and the paragraph is added to genus *Delphinus,* as nr. 28.
The Senegal Dolphin seen in great numbers along the westcoast of North Africa inside the 100 fathom line, can hardly belong to an unknown species. Published photographs of *Stenella frontalis* caught or beached in that area indicate that it could belong to that species or at least to the genus *Stenella.* The demarcation between dorsal and ventral pigmentation however is different in the Atlantic Spotted Dolphin and the Senegal Dolphin. This is a rather vital field-mark and the status of unidentified is maintained. The description is given as nr. 47.
The Illigan Dolphin in shape, size and locality could be a *Peponocephala electra* (nr. 67). This species was still unknown to science when the author made his illustration of the "Illigan Dolphin", described as nr. 48.
The difference between the rather brightly coloured Illigan D. and the jetblack Little Killer is big enough to maintain this dolphin as unknown, until more material for comparison becomes available.

SENEGAL DOLPHIN, plate nr. 47.
species   *Stenella spec.?*

This dolphin was observed and recorded by the author off the Senegal coast, and in outward form and appearance could not be identified with any of the known species. The shape is rather similar to that of the Greek Dolphin described as nr. 40 and possibly a *Stenella spec.* or subspecies.

area	Tropical West African Coast (Chart 3).
sea temperature	25–30° C.
length	Approximately 6 feet (1.8 m).
weight	Estimated at 150–200 lbs (68–90 kg).
food	Probably Fish and perhaps Squid.
teeth	No record.
illustration	From life. Smudgy brown above, sharply but not very distinctively separated from the dirty white undersides.
special features	A short plump dolphin with a rather short and stout snout. They are very playful, often attracted to ships where they play around the bow, alongside and especially in the spreading bow wave astern.
speed	Up to 18 knots.
breath	Not recorded.
schools	50–100, sometimes 5–10, but probably also in thousands.
biotope	Around the 100 fathom line.
immatures	Not observed.
notes	A regular visitor to ships between Dakar and 5° N, probably also further east.
	During the winter the same area is crowded with Common Dolphins whose bright colouring is much faded in that season. The two "species" are difficult to distinguish at a distance at that time especially when in large schools. This might be the reason that so little is known about the Senegal Dolphin. The species, however, is not quite unknown. The author has seen photographs of it in a journal (Cadenat.), but that does not give exact information about its taxonomy. Reckoning with the information available, it is not impossible that the Senegal Dolphin will turn out to be a *S. frontalis.*

ILLIGAN DOLPHIN, plate nr. 48.

species	*Peponocephala spec.?*

This dolphin was observed and recorded by the author on the north coast of Mindanao (Philippine Islands) and appearance up till now could not be identified with any of the known species.

area	North Coast Mindanao (Philippine Islands) (Chart 5).

sea temperature	28–30° C.
length	7–8 feet (2.1–2.4 m).
weight	Approximately 250 lbs estimated (113 kg).
food	Probably Squid and Fish (feeding with *Globicephala*).
teeth	No record.
illustration	From life. Several schools were seen at close quarters. Dorsally dark brown which, via orange, becomes yellow on the flanks. The underside is pink.
special features	A ploughshare head, similar to the Pacific White Sided Dolphin (*Lagenorhynchus obliquidens*) (nr. 21) or *Phocoena* (nr. 7), no snout was noticed. The dorsal fin is similar to that of *Delphinus delphis* (nr. 25).
	They paid no attention to the ship, but came quite close and when cruising, regularly jumped out of the water.
speed	Approximately 15 knots.
breath	Not recorded, estimated at every 5 to 15 seconds.
schools	10–30.
biotope	Coastal and deep water as typical for that area.
immatures	In May there were immatures but no newborn calves were noticed.
notes	Several schools of 10 to 30 animals were seen in Illigan Bay, off Aligay Islands. One group of 10 was in company with a school of 20 female and young Pilot Whales (nr. 65).
	Judging from form and locality this dolphin could be a Little Killer (*Peponocephala electra* nr. 67), which species was still unknown to science when the author made his illustration. Because of the difference in colouring, this separate paragraph will be maintained awaiting further observations.
	The common name is that of the locality where it was seen for the first time.

GROUP XI RIGHT WHALE DOLPHINS
Genus: *Lissodelphis* Gloger, 1841.

This genus has two species and is very specialized because they have no dorsal fin. One of the functions of this fin must be a stabilizing effect when swimming under water, and it developed in these mammals as it did in fish.

The only other *Cetacea* without one are the Right Whales. Their swimming is mainly a matter of going up and down between food and air.

In the slower species such as Narwhal, White Whale, the easy-going Fresh Water Dolphins and River Dolphins *Neophocaena, Sousa,* and *Orcaella,* the fin is reduced to a ridge.

Another possible function of the dorsal fin is to give its owner, fish (sharks) or whales, its exact position under water when swimming near the surface.

It is therefore doubly remarkable that this genus, of fast and oceanic species did not develop one, the flukes are also comparatively small.

Dr. Fraser remarked that these animals perhaps found compensation in body form. All dolphins have a greater height than width, except this genus, and the same can be said of the big Finner Whales, who, in comparison, have only a tiny dorsal fin.

The real shape of these large Whales is, however, difficult to judge because we can only get a good look at them when they are "high and dry" and then, of course, very much misshapen.

species	NORTHERN RIGHT WHALE DOLPHIN or PEALE'S DOLPHIN, plate nr. 49. *Lissodelphis borealis* (Peale, 1848).
area	North Pacific Ocean and Behring Sea. On the eastern side down to the Santa Catalina Islands, California, and on the western side to the east coast of Honshu (Chart 5).
sea temperature	5–13° C.
length	7–8 feet (2.1–2.4 m).
weight	150–180 lbs; 23 lbs/ft (68–90 kg).
food	Squid and Fish.
teeth	2 × 44 in upper jaw, 2 × 47 in lower jaw; small, sharp and pointed.
illustration	From life and photograph. The dolphin is seen from the underside to show the specially shaped white area on the

belly. Other dolphins who have a similar "lozenge" marking are *Globicephala*, Grampus, and in some cases, the three smaller killers.

special features	A slim and beautifully shaped dolphin, like a drop of liquid. The absence of the dorsal fin is striking, and gives it a likeness to seals. It is faster and more graceful. The beak is fine and slender. The width of the body is greater than the height.
speed	15–17 knots.
breath	Not recorded. As a deep water species it can probably stay underwater for an extended period (minutes).
schools	20, to several hundred in the northern part of their area.
biotope	Marine, cold water and deep ocean. They are rarely seen on coastal shipping routes.
migration	Although they have been recorded in the southernmost part of the area throughout the year, there is very likely a migratory movement north in the spring and south in the autumn.
immatures	Newborn calves have been recorded in April and May; length 2 feet (which is very small – 0.60 m) and weight 15 lbs (6–8 kg).
notes	The schools associate with Pilot Whales (nr. 65) according to Norris and Prescott. The author met hundreds coming in to Los Angeles on an easterly course from the Pacific on July 7th, 1958, passing south of the Santa Cruz Islands. All schools were combined with larger schools of *Lagenorhynchus obliquidens* (nr. 21).

The first group was of 20 Peales Dolphins 120 miles west of St. Michael. Over the last 100 miles to this island, he recorded about 300 Peales Dolphins and 1000 *Lagenorhynchus obliquidens*.

The Peales Dolphins came no further, but the latter species continued south and east of these islands.

SOUTHERN RIGHT WHALE DOLPHIN or PERON'S DOLPHIN, plate nr. 50.

species	*Lissodelphis peroni* (Lacépède, 1804).
area	Southern oceans – sub-antarctic (Chart 5).
sea temperature	5–15° C.
length	6 feet (1.8 m).

weight	Approximately 150 lbs (68 kg).
food	Squid and Fish.
teeth	4 × 43, small, sharp and pointed.
illustration	From photographs (publication by Fraser, 1955).
special features	No dorsal fin and the contrasting black back against the white body and snout make it very easy to recognize. The beak appears to be stouter than in the northern species.
speed	Approximately 15 knots estimated.
breath	Not recorded.
schools	2 or 3 to probably 10 and 20.
biotope	Marine and oceanic, sub-antarctic and deep water.
immatures	Not recorded.
notes	It is known to associate with schools of the genus *Lagenorhynchus* but the schools do not mix.

A New Zealand publication by Gaskin (1968) mentions reports of large schools ranging from 200 to 1000 eastward of South Island roughly along the 100 fathom line, during January 1964. These numbers seem to be out of tune with the rarity of the species. A school of 500 to 1000 was reported and filmed 200 miles of St. Pauls Isl. (Ind. Ocean) on March 24, 1968 by the captain of the British MS "Auckland Star".

The only other 3 records during the last 19 years came from Mr. K. Salwegter. A school of 5 about 250 miles SSE of Gough Island in the Sth. Atlantic and a school of 6 about 100 miles off shore near the SW corner of Australia, and one of 50 halfway the Australian Bight, both in August (1968 and 1970).

ROUGH TOOTHED DOLPHINS
Genus: *Steno* Gray, 1846.

The dolphins in this group form a separate genus because the conical teeth have a rough surface above the gum which can be felt when scratching it with a fingernail. There is officially one species: *Steno bredanensis* (Lesson, 1828). When, in 1950 the author started his studies on dolphins and whales he used a publication, the book about "Mammals of the Netherlands" of Ysseling & Scheygrond and a "Guide to Whales and Dolphins" of the British Museum of Natural History in London, which were available in the Library of the University of Utrecht.

Both publications described a small, roughtoothed dolphin as *Steno perniger* – Elliot's Dolphin – living in the tropical parts of the Indian Ocean. From description and illustration the author recognized it as a dolphin he had often encountered in large and shy schools between the Gulf of Aden and Ceylon. The author has never seen another publication concerning this species. The writer knows this dolphin however well enough and because neither in size, nor in habits it resembles *Steno bredanensis,* he gave this small dolphin a separate paragraph using the discarded name *S. perniger* as indication.

In 1936, as quoted in the Marine Mammals List, Mirando Ribeiro described a new genus and species – *Stenopontistes zambesicus* from this area. It is mentioned in the list under the genus *Steno* without details.

Perhaps this is the same dolphin. Because it is almost purely oceanic it is very well possible that it has thus far escaped the notice of science. Really oceanic species are seldom found stranded, apart from the fact that strandings in that area for an estimated 99 % will remain unreported.

	ROUGH TOOTHED DOLPHIN, plate nr. 51.
species	*Steno bredanensis* (Lesson, 1828).
	(= *S. rostratus*, Desmarest 1817 preocc.).
area	Tropical and sub-tropical oceans (Chart 5).
sea temperature	24–30° C.
length	8 feet (2.4 m).
weight	Approximately 300 to 350 lbs, estimated 40 lbs/ft (136–158 kg).
food	Not recorded.

teeth	4 $\times$ 20 to 27, conical and the exposed surface rough not smooth.
illustration	From life and illustration.
special features	The peculiar shape of the head (like a pike), with the snout laterally compressed; the purplish body colour with many white starshaped marks and the white beak.
	This dolphin comes to the bow to play, leaping with flat curves out of the water to reach that position.
speed	Over 15 knots.
breath	Every 6 to 7 seconds.
schools	3 and 4 to 8, perhaps more.
biotope	Deep water, preferably close to the coast. The author encountered it in the Philippine Islands twice, in the Cape Verde Passage, and once east of Masbate.
immatures	Newborn calves of $\frac{1}{3}$ length were observed in the Cape Verde Passage on 20th June.
notes	Very little is known about this dolphin (cf. Fraser, Kellogg) but it is known to be spread out over the whole tropical Atlantic. Next to the places mentioned in the Philippines, Mr. K. Salwegter recorded it on the east coast of Australia, from 30′ south to 60′ north of latitude 30°. The great number of very old skulls in Dutch museums indicate that in the last century mass strandings must have occurred in Indonesia.
	Recent (1965) mass strandings occurred in Florida and in Senegal. Temperate records (Holland, Brest, Mosselbay) are single strandings or only skulls.
	One specimen was caught in the Hawaiian Islands and transported alive to the "Sea Life Park" on Oahu Island, more have been recorded there since.

ELLIOT'S DOLPHIN?, plate nr. 52.

species	*Steno perniger* (Blyth, 1848).
area	Western Arabian Sea and Indian Ocean (Chart 5).
sea temperature	27–32° C.
length	4–5 feet (1.2–1.5 m).
weight	Approximately 110 lbs estimated (50 kg).
food	Not recorded.
teeth	Not recorded.

illustration	From life and illustration.
special features	A small and very shy dolphin; dark brown-black above, the underside red with dark spots.
	The snout is less conspicuously separated from the forehead than in other dolphins.
	They are always afraid of a ship and try to swim away from it. The school closes ranks till they are almost touching each other and start jumping in long rows of 10 and more together, returning in the water at an angle of about 45°, which clearly shows the belly.
	Large schools have the habit of appearing "from nowhere", the sea is smooth and empty and suddenly it is alive with hundreds. Whether they come from great depths or just escape notice because they keep quiet, the author has not been able to discover.
speed	11 knots maximum.
breath	Every 3 seconds, the whole school is on the surface for one minute, then it makes a deep dive for $1\frac{1}{2}$ to 2 minutes. When cruising every 3 to 7 seconds.
schools	100–1000, and probably more; an average school is 300.
biotope	Deep water, often oceanic, sometimes off the 100 fathom line.
immatures	Calves of all sizes are seen all the year round, but more in the first half of the year.

SUBFAMILY DELPHININAE

GROUP XII   RIDGEBACKED DOLPHINS
Genus: *Sotalia* Gray, 1866 and
Genus: *Sousa* Gray, 1866.

The original genus *Sotalia* was in 1968 divided into two genera: *Sousa* in Asia and Africa and *Sotalia* in Sth. America.
They have a worldwide distribution in coastal areas, tropical and sub-tropical climates. They are also called RIVER DOLPHINS or ESTUARY DOLPHINS.
Sousa plumbea, which lives round the coasts of the Western Indian Ocean, including the Persian Gulf and the Red Sea, is the typical representative of the group, and in the geographical centre of distribution. Eastward from there live three species:
– *Sousa lentiginosa;* round the Bay of Bengal to Malaya;
– *Sousa borneensis;* from Malaya and Borneo to Thailand; and
– *Sousa chinensis;* from Thailand to China.
Westward from the centre, the Atlantic Ocean also three species:
– *Sousa teuszi; al*ong the coasts of tropical West Africa;
– *Sotalia guianensis;* from Rio de Janeiro to Guiana, and
– *Sotalia pallida;* (= *S. fluviatilis*) on the Amazon River.
These dolphins have long slender bodies and in *Sousa* a small dorsal fin situated on a ridge varying in length from one foot to almost $\frac{3}{4}$ length of the body. The snout is remarkable long, thin and rodlike.
They are normally not frisky and rarely jump out of the water, quietly break surface to breathe and equally quietly disappear again, usually paying no attention to ships. Although they belong to the family of Ocean Dolphins, they are never seen at sea and certainly not in the ocean. From his personal observation the author concludes that these dolphins stay within one or two miles from the coasts they inhabit.
They are typical Estuary Dolphins and perhaps as well known to seamen as their more spectacular brothers in the open oceans, because they can be regularly seen when entering harbours such as Cochin, Bombay, Bahrein, on the Shatt al Arab, Rastanura and occasionally in the Suez Canal.
Only the last mentioned species is a pure fresh-water river dolphin, the others live inshore and go up as far as the flood tide brings the salt water.
From the centre eastwards there is a gradual physical change from the large lead coloured *S. plumbea* with a long conspicuous ridge, to the smaller pure white or pink *S. chinensis* with a much smaller ridge. There is even a con-

troversy as to whether the last two species are not only one. The sea area in that direction is, of course, continuous and the different species touch and sometimes overlap each other.

In westward direction the physical change is greater. The homewaters are first separated by the cold Benguella Current and next by the Atlantic Ocean. The W.-African species are still *Sousa* types (although the shape of the tail reminds the author of the *Platanistidae*). They are imperfectly known. The original description was of a damaged specimen, mutilated by sharks in a landlocked lagoon, and incorrect, including the fact that it fed on vegetable matter. In 1866 all "Estuary dolphins" were brought together in the genus *Sotalia*. The two South American species are very similar in form but different from the rest.

In the field they could be more easily mistaken for a small kind of *Tursiops*, missing the long snout and the dorsal ridge, they have a normal triangular, rather large dorsal fin.

For this reason the group was divided in 1968 in Genus *Sotalia* for the Sth. American animals and Genus *Sousa* for the Afro-Asian populations.

The number of teeth is for both genera about the same, more than in *Tursiops*, less than in *Delphinus* and *Stenella*.

	AMAZONE RIVER DOLPHIN or BUFEO, plate nr. 53.
species	*Sotalia pallida* Gervais, 1855.
	[= *S. tucuxi* Gray, 1856].
	[= *S. fluviatilis* Gervais et De Ville, 1883].
area	Amazone River and tributaries to 1600 miles from the sea (Chart 5).
water temp.	26–30° C.
length	4–6 feet (1.2–1.8 m).
weight	70–90 lbs (32–41 kg).
food	Fish.
teeth	4 × 28 to 31.
illustration	From photographs and description. Bluish to pearl grey above, darker forward. The flipper has the dorsal colour on both sides. Lower jaw and throat are pinkish white and the rest of the underside is white. The animals probably become lighter as they get older.
special features	They share the whole area with the Amazone Dolphin or Boto (nr. 2) and the schools are often seen together. The Bufeo is smaller, quieter and more difficult to detect, sliding

to the surface to give a little puff, showing head, back and dorsal fin, then diving rather steeply with a strong arching of the tailstock. There are usually half the numbers of Bufeo compared to Botos in the same area.

They prefer the rivers above the lakes and lagoons, but are seen in both, remaining close inshore, also entering inundated forests, whereas the Boto prefers the lakes and lagoons and stays more midstream.

Only young calves were noticed to be interested in boats, adults were shy.

speed	Cruising about 2 knots, capable of 5 to 6 knots.
	Their movements are quicker than those of Botos.
breath	Ranging from every 5 to 85 seconds.
schools	Small and spread out. An estimate was made at 2 to 3 per sq. mile.
biotope	See special features.
immatures	Calves were seen during February and March, but there are no observations in the other months. They were black above and pink below, the two colours being sharply separated.
notes	They are held in high esteem by the surrounding inhabitants and are not hunted as food or for their oil.

GUIANA RIVER DOLPHIN, plate nr. 54.

species	*Sotalia guianensis* (P. J. van Beneden, 1864).
	[= *S. brasiliensis* E. van Beneden, 1875].
area	North-east coasts of South America from Rio de Janeiro to Guiana. The distribution perhaps includes Venezuela. The author received several reports of this "type" of dolphins from Maracaibo (Chart 5).
sea temperature	24–30° C.
length	5–6 feet (1.5–1.8 m).
weight	Estimated at 100 lbs (45 kg).
food	Fish.
teeth	*S. guianensis* 4 × 28/31, *S. brasiliensis* 4 × 33/34–31/35.
illustration	From frozen laboratory specimen, and description.
	They are very similar to the Bufeo, but are slightly larger and have a darker colour.
special features	Swimming and surfacing in the typical *Sotalia* way. The rather long triangular dorsal fin is different from that of

other dolphins.

speed	Probably 5–6 knots; when cruising, 2–3 knots.
breath	Not recorded.
schools	2–20, and more.
biotope	Coastal rivers and in shore in the area, salt and brackish water.
immatures	Not recorded.
notes	It is common in the Bay of Rio de Janeiro. By the local, rural population it is considered a sacred animal and it is not hunted.

CAMERUN RIVER DOLPHIN, plate nr. 55.

species	*Sousa* (originally *Sotalia*) *teuszi* (Kükenthal, 1892).
area	Tropical West Africa from Camerun to Senegal (Chart 5).
sea temperature	20–30° C.
length	± 7 feet (2.2 m).
weight	± 190 lbs (85 kg).
teeth	Upper jaw 2 × 29 to 30; lower jaw 2 × 27; small, open spaces between the teeth.
food	Fish.
illustration	From photographs. The dorsal ridge is probably less conspicuous in real life.
special features	No records.
speed	Cruising, probably 2 knots; capable of up to 5 and 6 knots.
breath	Probably ranging from every 5 seconds to a maximum deep dive of 2 minutes.
schools	Probably small and spread out.
biotope	River entrances and inshore coasts in the area.
immatures	No records.
notes	Very little is known about this dolphin. Until recently the damaged type-specimen of 1892 was the only animal available, described by Kükenthal with the information he received from the finder, Mr. Teusz. Since 1956 more stranded animals have been found, all near Dakar, and scientifically described by Cadenat. A living dolphin was recently obtained.

LEADCOLOURED DOLPHIN or MALABAR DOLPHIN, plate nr. 56.

species	*Sousa* (originally *Sotalia*) *plumbea* (G. Cuvier, 1829).
area	Indian Ocean coasts, from Cochin to Port Elisabeth S.A., including the Persian Gulf and the Red Sea (Chart 5).
sea temperature	18–30° C.
length	Up to 8 and 9 feet (2.4–2.8 m).
weight	Not recorded, estimated at about 200 lbs (90 kg).
food	Fish.
teeth	4×34 to 38.
illustration	From life and photographs. The grey colour appears rather light due to the numerous little white "spatters" all over the body. In the picture these are painted too large. The snout has a diameter of about 2″. The lower jaw is mostly white and also the anterior part of the upperjaw, the posterior margin of the flukes and sometimes the little dorsal fin.
special features	This is a typical *Sousa* with a large dorsal ridge, in some older animals it runs from halfway blowhole-dorsal fin to the flukes. This ridge mainly shows when the animal is in a straight, horizontal position; when the back is arched for diving it dissolves into the body. They surface slowly with a 30° angle, exposing snout, then, levelling of, the head and back as far as the dorsal fin, followed by the arched tail-stock, diving again at a rather steep angle. They rarely jump and seem to be more frisky during darkness than in daylight. When sounding, the flukes sometimes show.
speed	Cruising, 2 to 3 knots, with an estimated maximum of 6 knots.
breath	After breathing about 4 times every 8 seconds, a dive lasting from 20 to 40 seconds. The longest dive the author recorded lasted two minutes.
schools	2–20 or more, spread out.
biotope	Inshore on coasts and up rivers in the area. Salt or brackish water, preferably in depths of 2 to 6 fathoms with strong tidal currents. Always swimming against the tide.
immatures	Newborn calves were seen in March and April.
notes	When looking for this dolphin, a sharp lookout should be kept on the surface of the water. They are not attracted by ships, but neither are they afraid of them, and sometimes come up only a few yards from the

hull. Rarely swim before the bow (speed of less than 4 knots).

In anchorages they catch fish, hiding or resting in the shadow of the ship during the day, diving on one side and coming up with the fish on the other side.

The ship's agent in Cochin, who lives right on the water's edge in the entrance channel from sea into the lagoons, informed the author that on several nights he was woken up by the noise of their continuous leaping and splashing under his window, around the time of the full moon.

Some of the better know places where they can be observed are: Cochin, Bombay, around Bahrein and Ras Tanura, Kuwait, Shatt al Arab, Khoramshar, Aden, Djibouti, Suez Roads, and the Bitter Lakes in the Suez Canal.

The author and his crew hunted them to obtain a specimen for science in the Persian Gulf for over a week in two lifeboats with harpoons, but although there were scores around, none ever surfaced near enough to be hit. This occurred in clear and muddy water which indicates that their sonar is probably well developed.

In the Bitter Lakes they remain also during the winter, when the seawater is 18° C, which is rather important considering the possibility of this species reaching the Atlantic Ocean around South Africa. In Djibouti the colour is light grey; in the Bitter Lakes, black grey.

Recently several have been caught in Seine nets, both in Durban Bay and Algoa Bay (Port Elizabeth). Some zoologists consider this species and the next as one.

SPECKLED DOLPHIN (see notes), plate nr. 57.

species	*Sousa* (originally *Sotalia*) *lentiginosa* Owen 1866.
area	Bay of Bengal and Malacca Straits to Klang River (Chart 5).
sea temperature	28–30° C.
length	Approximately 6 feet (1.8 m).
weight	Estimated at about 120 lbs (51 kg).
food	Fish.
teeth	4×33.
illustration	From life. The illustration was made in the Malacca Straits on the Deli and Klang Rivers.

This dolphin is neither a *S. plumbea*, nor a *S. borneensis*, but somewhere in between, having an overall grey colour, but considerably blotched with whitish and pink areas also a much shorter dorsal ridge.

Only half-grown immatures were seen, jumping clear of the water, these were very light grey and had pinkish white bellies.

The author has no proof but presumes he portrayed the *S. lentiginosa.*

special features	The typical *Sousa* rod-like snout and behaviour.
	They can usually be seen on the Deli River somewhere between the outer bar and the Harbour office of Belawan, which are 10 miles apart, swimming against the tide. *Orcaella* (nr. 71) are often in the same area.
speed	2–6 knots; the tidal currents reach a velocity of 4½ knots.
breath	Every 20 to 25 seconds, sometimes longer.
schools	2–20.
biotope	Muddy rivers and coasts, depths of 20 to 40 feet, salt and brackish water. The author has never observed them at sea outside the bar.
immatures	No record. The data the immatures were seen was not recorded.
notes	This dolphin has not been observed further south than the shallow banks stretching across Malacca Strait near Port Swettenham.
	The Riou and Lingga Archipelago further south is the territory of the *S. borneensis*. This genus does not occur around Banka Island. The much frequented Musi River (Palembang, Pladjoe (Shell) and Sungei Gerong (Stanvac), has no dolphins at all.
	The author has not seen this dolphin in or near the entrances to the Hoogly River (Calcutta), and Karnaphuli River (Chittagong) but it must be borne in mind that these estuaries are fresh water. They were observed on the bar of Pussur River.
special note	On the 21st April, 1960, the author had the opportunity to observe at close quarters a "freak dolphin", which he indicated in his notes as a "Silvergrey *Tursiops*", whilst anchored at Deep Water Point off Port Swettenham, in depths of from 8 to 12 fathoms.

This dolphin was about 7 or feet long, silver grey above, which via a blotchy area changed to white on the flanks and the underside. Behind the dorsal fin the colour was pale pearl grey and in front of it were some darker patches. The snout was long and rod-like as in *Sousa*, the lower jaw freckled white. The dorsal fin was big and firm, shaped like that of *Tursiops*, and there was no dorsal ridge. The flukes were rather triangular as in Sperm whales.

It played around the anchored ship for about ten minutes, turning round in small circles, on or just under the surface, coming up frequently, each time with the beak wide open. The inside of the mouth was pink.

It swam several times with the belly upwards, made deep dives of approximately one minute and jumped clear of the water only once.

The author could not identify this dolphin with any of the species or forms known to him.

If it is at all possible this animal looked like a cross between a *S. plumbea* and a *Tursiops* of the Indian Ocean (see notes nr. 58).

BORNEO WHITE DOLPHIN, plate nr. 58.

species	*Sousa* (originally *Sotalia*) *borneensis* (Lydekker, 1901).
area	Riou Archipelego, east coast Malaya to head of the Gulf of Siam, the Serawak coast of North Borneo, perhaps further along the coasts of this island (Chart 5).
sea temperature	28–31° C.
length	5–6 feet (1.5–1.8 m).
weight	Estimated at 120 lbs (54 kg).
food	Fish.
teeth	4×35 to 33.
illustration	From life and photographs. The average colour is ivory or dirty white. The eyes are dark. Light grey specimens also occur.
special features	*Sousa* form and behaviour. The white colour is easily recognized. When sounding they occasionally show the flukes.
speed	4 to 5 knots.
breath	Several times in quick succession every 2, 3 to 5 seconds, then a deep dive of 65 seconds.

schools	2–4, sometimes up to 10, but no more. Spread out. Only pairs swim close together.
biotope	Rivers and coasts in the area, salt and brackish water. In the Rajang River (Serawak) they come to 20 or 25 miles upstream.
immatures	No records.
notes	In May, 1965, when leaving Singapore Strait in an easterly direction, the author met one pair, swimming at a steady pace on the surface from the direction of the Outer Reefs northward of this strait towards Horsburgh Lighthouse. One was light grey, the other from snout to flukes pure sugar pink. These *Sousa borneensis* are the only ones he has ever seen "at sea". Several dolphins which are as far as could be judged identical with the Borneo White Dolphin live in captivity in the Coolangata Oceanarium of Mr. Jack Evans (40 miles south of Brisbane, Australia).
	These dolphins are caught locally in the North Australian River estuaries and are larger than the local *Tursiops* in the same oceanarium.
	The type of biotope of North Australia is very much like that of North Borneo, characteristic for the *S. borneensis*. The author wonders whether these dolphins are one and the same species or should be classified as a separate species or subspecies as Australian White Dolphin.
	See also notes on "freak" dolphin nr. 57.

	CHINESE WHITE DOLPHIN, plate nr. 59.
species	*Sousa* (originally *Sotalia*) *chinensis* (Osbeck, 1765). [= *S. sinensis* (Desmarest, 1822)].
area	From the Gulf of Siam, northward along the coasts of Vietnam and China (Chart 5).
sea temperature	20–28° C, perhaps colder.
length	Approximately 6 feet (1.8 m).
weight	Estimated at approximately 120 lbs (54 kg).
teeth	4×32.
illustration	From description. There are several records of pure pink animals and some authorities consider it conspecific with the *S. borneensis*. The pure pink dolphin which the author observed in Singapore Strait would confirm this.

special features	*Sousa* form and behaviour.
breath	Every 3 to 5 seconds with deep dives probably up to one minute.
speed	4 to 5 knots.
schools	Less than 10.
biotope	Inshore coasts and river entrances in the area.
immatures	No records.
notes	This dolphin is not very well known. Many colleagues of the author who frequent the coasts and rivers in the area have never seen it except in the Gulf of Siam, Bangkok, in the Canton River and Swatow, but it has not been reported in Hong Kong, the Mekong (Saigon) or Song Koi (Hanoi).

It very likely migrates to warmer climates in the winter, the temperatures of the seawater from Hong Kong northward are too cold for a tropical species. From January to April the isotherm of 15° C comes down as far as Amoy, whilst during the summer temperatures reach 20° and over as far as Korea.

GROUP XIII   BOTTLENOSED DOLPHINS
Genus: *Tursiops* Gervais, 1855.

The dolphins of this genus can be considered the best known of all whales. Their distribution covers practically the whole tropical and temperate world. They are numerous, live in considerable schools near the coasts and along many main shipping routes, well into shallow water off harbour and river entrances. They are friendly and playful; when a ship comes in the neighbourhood of a school, often a dozen to a score of animals will take time off to come and play before the bow, alongside and astern of the ship, approaching with a fantastic speed, taking great leaps through the air to reach the target. They often stay for half an hour or more, then gradually fall back but remain visible for a long time, jumping 10 to 15 feet in the air, out of the spreading bow wave astern making saltos in every direction, causing splashes which can be heard over a mile away.

Their intelligence is almost common knowledge at the present day through articles in magazines, films and TV, describing and showing wonderful cooperation in executing complicated tests or amazing acrobatics.

These tests continue in many seaquaria, oceanaria, dolphinaria, marinelands or sealife parks for the benefit of science as well as for the enjoyment of the public. F.i. Dr. W. H. Dudok van Heel at Harderwijk-Netherlands; Dr. William Schevill – Oceanographic Institute, Woods Hole, Mass; Dr. Norris and Prescott – Marineland, California; Dr. J. C. Lilly – Marineland, Florida; Sea World – S. Diego, California; Mr. and Mrs. Prior at Sealife Park, Hawaii.

Since mythological times dolphins have been reported to seek out and play with men. A recent recurrence in the New Zealand village of Oponene has again proved this to be true (cf. "Dolphins" by Anthony Alpers). As a very socially minded herd-animal it likes company. New Zealand is only occasionally reached by schools from further north. The young specimen was solitary and obviously lonely, seeking out the company of people bathing and swimming off the beach to play with and had a special affection for one little girl.

The famous Pelorus Jack, which daily accompanied the ferry from Wellington to Nelson through Pelorus Sound for 30 years, was probably also a *Tursiops* and not a *Grampus* as originally stated.

Before World War II children on the "Plage des Enfants" eastward of the harbour of Port Said, were known to be able to touch the *Tursiops* playing and cruising in the same area, and in Florida USA where the first Marineland opened its doors many years ago, several families now have their own pet dolphin playing in the water round the house.

The U.S. Navy has harnessed dolphins in training, passing messages to divers and learning to help in rescue operations. This rescue sense is a natural instinct. The dolphin is one of the few herd-animals which will stand by and help a handicapped or wounded herd-member, supporting it and if necessary assisting it to the surface for breathing, until it is dead or cured.

Scientists are trying to teach the dolphin to use that instinct to help them, but are even more eager to learn how their perfect Sonar works, and the secrets of the wonderful shape and skin which nature has perfected for speed and economical use of power under water.

Ships with a dolphin snout, the bulb, on their bows, have surprised everybody by making better speeds with a considerably lower fuel consumption.

The supple, smooth and flexible skin will be far more difficult to imitate, but who knows?

The division in species of the genus *Tursiops* is still a point in discussion amongst the taxonomists.

The nominate race: – *Tursiops truncatus truncatus* lives in all oceans from the east coasts of both Americas (more grey), Western Europe and Africa (more brown), Mediterranean, Suez Canal, perhaps also Red Sea and Indian Ocean (lighter brown) to Japan and Australia in the Western Pacific Ocean (dark grey), occasionally to New Zealand.

Tursiops truncatus gilli, almost black but with sepia brown in the colour, is a sub-species off the west coast of North and Middle America.

The Indian Ocean and Australian populations were formerly called *Tursiops catalania,* Gray 1862, and by some zoologists are now considered to be *Tursiops aduncus,* Ehrenberg 1833. This last name, however, is originally given to the Red Sea Dolphin, a small species described as a *Tursiops,* exclusively living in the Red Sea and Gulf of Aden, greenish in colour with dark undersides and a thinner snout. The author is of the opinion that this is a different animal and described it for that reason in a separate paragraph (nr. 63). The exact place of the Red Sea Dolphin in the family of the *Delphinidae* is one of the problems which still has to be solved.

Tursiops nuuanu, which by some zoologists is considered conspecific with *T.t. gilli,* is a small, dark brown and shy dolphin, mainly oceanic, living in enormous schools in the eastern tropical Pacific.

Tursiops are extremely rare in South Africa. In the literature only one skeleton is mentioned. The author has, however, seen a considerable number of

Tursiops dolphins in Walvis Bay, in the bay and in the harbour. They were big, black brutes, full of scars who, besides their ugliness, accompanied the ship and tug only on either side of the propeller. One had an enormous scar where obviously a propeller had hit it right across the back. Whether these are Bottle-nosed Dolphins belonging to the West African population or a different local race, the author cannot judge. The sea water temperature there is 15° C.

BOTTLENOSED DOLPHIN of the Atlantic nr. 60, of the Indian Ocean nr. 60a, of S. Atlantic nr. 60 b, plate nrs. 60, 60a, 60b.

species	*Tursiops truncatus truncatus* Montagu, 1821.
	[= *T. catalania* Gray, 1862].
	[= *Sotalia gadamu* Owen, 1866].
	[= *T. gephyreus* Lahille, 1908].
area	World wide tropical to temperate (Chart 6).
sea temperature	Above 15° C.
length	10–12 feet (2.5–3.7 m).
weight	Up to 600 and 800 lbs; 60–70 lbs/ft (270–370 kg).
food	Fish and Squid (30–65 lbs/day).

Big fish are "softened" by smashing them on the water or rubbing them on the sea floor. The chalky "shell" of the squid is discarded when coming to the surface. *Tursiops* kill sharks which threaten their young, by ramming and butting.

teeth   4 × 20 to 22, rather large, diameter $\frac{1}{3}$–$\frac{1}{2}$" (8–12 mm).

illustration   From life and photographs. The basic colour is very dark grey with a brownish tint. The amount of light, especially direct sunshine, has influence on the observation. The West Atlantic and West Pacific populations seem to be smaller and more grey, the European and Indian Ocean specimens larger and more brownish. In the tropics, nearly all dolphins have a darker, oval shaped area on the back, where the skin is regularly exposed to the air when blowing. This is more accentuated in older animals (more scarred), than in younger and the author believes it is caused by sunburn. On a young mother, dark chocolate-brown, in the Persian Gulf, it was hardly noticeable. The South Atlantic population in the Benguella Current is almost black, big and coarse in appearance and full of scars (nr. 60b).

special features	The lower jaw protrudes in front of the upper jaw. When the animals are observed from above, which is often quite easy because they always come to the bow of ships, swim, roll and jump there, this gives the impression that they have a "White Nose". They are very playful and from one or two as many as twenty manage to keep station in front of the bow, jumping, and rolling. Some individuals keep this up for over 15 minutes, before retiring and continuing the game alongside and astern. The dorsal fin has a characteristic shape.
speed	Cruising, 5 to 6, normal, 13 to 18, and capable of over 20 knots.
breath	Varying from 6 to 30 seconds, maximum time onder water 7 minutes. Speed 13 knots, 15–20 seconds; 16 knots, 10 seconds; 18 knots, 5–10 seconds. Mother with newborn calf: 4 $\times$ on the surface, then away 2 minutes, covering 200 yards.
schools	10–20, also 200 to over 1000.
biotope	On the continental shelf, inside the 100 fathom line, only rarely outside. Sometimes far up rivers f.i. 45′ up the Mississippi, but not in fresh water.
immatures	Gestation period 12 months. Newborn calves are 3 to 4 feet (0.9–1.2 m), weight 70 lbs (32 kg); when 7 feet reach 270 lbs (40 lbs/ft). From observations it seems that in the Atlantic they are born during autumn in the Indian Ocean during February and March.
migration	There is undoubtedly a migration in the Atlantic, northward during the spring, reaching northern Norway in the summer; southward at the end of autumn, one has to bear in mind that the sea water temperatures lag almost a season behind: still high at the end of October and still low during April and May.
notes	This popular dolphin is easily trained and is the backbone of most sea aquaria in the world. An interesting book on tests and trials with these dolphins is "Man and Dolphin" by John C. Lilly. Comparison with a seal roughly the same size gives the following figures. (Weddell seal, *Leptonychotes weddelli*, Lesson 1826). National Geographic Magazine, January 1966.

Length: 9–10 ½ feet (2.7–3.2 m). Weight: 800/1000 lbs 90–100 lbs/ft. Newborn calves length 5 feet, weight 60/70 lbs, 13 lbs/ft. Breath for 2 to 3 minutes on the surface, can stay under water for half an hour. Dives to 250 fathoms.

PACIFIC BOTTLENOSED DOLPHIN, plate nr. 61.

subspecies	*Tursiops truncatus gilli* Dall, 1873.
area	Eastern Pacific Coasts from Los Angeles to Panama, in the Hawaiian and Galapagos Islands (Chart 6).
sea temperature	15–30° C.
length	10–12 feet (2.5–3.7 m).
weight	600 to 800 lbs; 60/70 lbs per ft (270-370 kg).
food	Fish (smaller than one foot), Squid, also Shrimps and Crabs.
teeth	4 × 20-22, bigger and stouter than in *Tursiops t. truncatus* 0.7″, (18 mm).
illustration	From life and photographs. There is a difference between a much scarred inshore shallow water and a less scarred deep water population. Observed from a distance the animal looks all black, but close by, under the bow it turns out to be dark sooty brown.
	The throat and middle line over the belly are lighter, round the anus is a round pink patch.
	These marks can only be seen when it rolls over or swims on its side.
special features	Typical *Tursiops* in form and behaviour. They nearly always come to the bow of passing ships to play.
speed	Estimated over 20 knots. The fat Galapagos population is slower.
breath	About 3 times ev. 6 to 10 sec., then a deep dive from 1 to 2 min.
schools	3 to 40, rather spread out, sometimes 150 to 200.
	The greatest concentrations were observed round Nicoya Peninsula in the summer.
biotope	Coastal in the area, occasionally marine. On both side of the 100 fathom line. The continental shelf along this coast is very narrow.
migration	There probably is a migratory movement in the northern part of the area.
immatures	Calves are born in May.

notes The author's recordings show a maximum distance off the coast of 200 miles and it was not observed during several crossings through and past the Revilla Gigedo Islands. This is the opposite of what he recorded for the smaller *Tursiops nuuanu,* who are practically pelagic and live in schools of hundreds close together.

Some *Tursiops* were seen in the Hawaiian Islands, but there was no opportunity to ascertain whether these were *T.t. gilli.* The colour was definitely lighter than that of eastern Pacific continental species.

The Galapagos population is very dark grey, and seems to have very short beaks. They are abundant inside the perimeter of the group (many hundreds) but rare outside.

LITTLE BOTTLENOSED DOLPHIN, plate nr. 62.

species	*Tursiops nuuanu* Andrews, 1911.
area	Tropical Eastern Pacific Ocean (Chart 6).
sea temperature	27–31° C.
length	7 to $7\frac{1}{2}$ feet (2.2 m).
weight	285 lbs (129 kg).
food	Fish and Squid.
teeth	$4 \times$ 20-22 about $\frac{3}{4}$ size of those of *Tursiops truncatus gilli* 0.5", (13 mm).
illustration	Life.
special features	Typical *Tursiops* form, usually living in large schools. They are afraid of, or at best indifferent to ships and swim away at right angles bringing the "danger" straight astern. Their jumping is with flat smooth curves and they sometimes rush along under the surface with only the blowhole and dorsal fin (10" = 25 cm high) exposed when blowing, similar to the Dall's Porpoise (nr. 12).
speed	± 15 knots.
breath	Not recorded (for dolphins which swim in large compact schools recording the breathing intervals is very difficult because it is impossible to follow one individual).
schools	Very large and compact, the animals swimming very close to each other, from 100 to 500 individuals is normal. Northeast of the Galapagos Islands the author encountered three schools of 100 and one of about 800 over a distance of 10

miles. Southeast of Acapulco a school was sighted estimated to be over 1000.

biotope　Oceanic and deep water, also up to the 100 fathom line near the coast. The type specimen is from 12° N and 120° W.

immatures　The schools contain all sizes from new born calves to fully grown immatures.

notes　This dolphin is by some authorities considered to be conspecific with the *Tursiops t. gilli*.
The differences noted by the author in the field, supported by the paragraph on this dolphin in the Puritan-American Museum Expedition's publication by Richard G. Van Gelder, have led him to describe it in a separate paragraph.

RED SEA DOLPHIN, plate nr. 63.

species　*Tursiops aduncus* (Ehrenberg, 1832)?
[= *T. abusalam* (Rüppell, 1842)].

area　Red Sea and Gulf of Aden (not in the Gulf of Suez) (Chart 8).

sea temperature　25 to 33° C.

length　± 6 feet (1.8 m).

weight　± 200 lbs (90 kg).

food　Fish and Squid.

teeth　2 × 26 upper jaw, 2 × 27 in lower jaw.

illustration　From life and photographs. The dolphin is dark, and as with all other dolphins the light, especially the direct sunlight, high, low or covered with clouds, makes a great difference for the observed colour. It has a definite olivegreen appearance.

special features　A smallish and slender dolphin with a dark underside. The snout has the projecting lower jaw, which gives the impression when seen from above, to be a white tip. The dorsal fin is similar to that of *Delphinus* in shape and not as in *Tursiops*.
They are very playful and nobody sailing the length of the Red Sea can miss them. Some, usually 10 to 20, come to the bow and alongside to play, often staying with the ship for 15 minutes. This species is one of the easiest to photograph.

speed　Cruising 3 to 4 knots, capable of speeds to over 20 knots.

breath	Every 15 to 30 sec.
schools	10 to 30 and 50, sometimes as many as 500 together. The smaller schools are rather compact.
biotope	Deep water, rarely inside the 100 fathom line.
	It is more abundant in the southern part of the Red Sea than in the north, goes right up to the head of the deep Gulf of Aqaba, but is never seen in the shallow Gulf of Suez.
immatures	All sizes are seen in the schools throughout the year.
notes	For this dolphin there is no common name available in the literature. The name Red Sea Dolphin seems appropriate. In the Marine Mammals List the species name *aduncus* is used to indicate all Indian Ocean and Australian *Tursiops* populations. Because The Red Sea Dolphin is so different in size, colour, the shape of the dorsal fin and dental formula from the larger Indian Ocean population of *Tursiops truncatus,* which, also occurs in the same area, the author has given it a separate paragraph.
	On the tailstock three light streaks from the lighter belly nearly always invade the dark dorsal greenish brown.

GROUP XIV   BLUNTHEADED DOLPHINS or KILLER GROUP

This group has no official name or status, all the members belong to different genera but share some striking features as the round blunt head or melon, a conspicuous dorsal fin and big strong teeth.
Some authorities have already considered to treat this group as closely related species and Dr. Nishiwaki of the Whale Research Institute of Tokio suggested the family name of *Globicephalidae*.

In most of the species, and perhaps all, the male is larger to considerably larger than the female. Two reach a length which brings them in the range of whales and in the common name this indication is mostly used.
Three of the known genera have "Killer" names:

> *Genus Pseudorca*   – False Killer
> *Genus Feresa*   – Pygmy Killer
> *Genus Orcinus*   – Killer Whale

Two other genera resemble them so much, especially in the immature stage, that in the field it is sometimes very difficult to recognize them.
These are:

> *Genus Grampus*   – Grampus
> *Genus Globicephala* – Pilot Whale

The 6th genus was previously classed as the rare *Lagenorhynchus electra* or Broadbeaked Dolphin from 8 skulls available. Recent strandings of living animals and subsequent publication of detailed information, revealed that this dolphin is very similar to the first two genera, especially in outward appearance. It was renamed:

> *Genus Peponocephala* – Little Killer is the name the author

would like to suggest instead of "broadbeaked dolphin" so far used for the practically unknown animal. Perhaps the Illigan Dolphin (nr. 48) belongs also to this group.

The 7th "Killer"-dolphin brought into this group is an animal which up till now does not belong to a known genus. It was several times observed by the author in the Gulf of Aden near the village of Alula, where the illustration was made.

There is a strong resemblance to the members of this group and until it can definitely be classed he suggests naming it the Alula Killer.

All known genera are deep water dolphins and have a worldwide distribution. Two belong to the rarest species, which until a few years ago were only known from skulls.

genus	GRAMPUS Gray, 1828.
	GRAMPUS OR RISSO'S DOLPHIN, plate nr. 64a, 64b.
species	*Grampus griseus* (G. Cuvier, 1812).
area	See Chart 1. It is more restricted than the general statement "All oceans tropical and temperate" indicate.
sea temperature	15–30° C.
length	12 to 13 feet (3.6–4.0 m).
weight	± 1500 lbs; 120 lbs/ft (680 kg).
food	Squid.
teeth	Upper jaw 0–2, Lower jaw 3 to 7 in the front part only. Diameter $\frac{1}{2}''$ to $\frac{3}{8}''$ (1.2 × 1.9 cm).
illustration	From life and photographs. The colour varies from practically uniform grey (nr. 64a) to almost white with numerous grey patches and scars all over the body (nr. 64b). The last part to become white is the area round the dorsal fin. The belly has an hourglass-shaped white marking more or less similar to that in the other members of this group and the *Lissodelphis* (nr. 49).
special features	The grey or whitish colour makes them conspicuous and with the high (1 foot 3″ = 40 cm) pointed dorsal fin easy to recognize. Their progress under water can be followed without much effort.
	The white specimens seem to be more common in the colder parts of their areas.
	They are not afraid of ships, but do not normally come and play. When a ship is heading straight for a school they alter their course a little and let the ship pass, then usually start jumping in the spreading bow wave $\frac{1}{2}$ mile to one mile behind the stern.
	Only once, near Djibouti did a school of 5 accompany the ship alongside at a distance of no more then 50 yards.
speed	15 knots, can maintain 17 knots for about 5 minutes.
breath	Normally every 6 seconds, often stays under water for

	several minutes and can remain submerged for $\frac{1}{2}$ hour.
schools	Always in small groups of 2 to 5, but with as many as 50 together.
biotope	Coastal in deep water, outside the 100 fathom line, occasionally marine and oceanic, probably when migrating.
migration	In all marked area's they occur throughout the year, but are more numerous round the north of Scotland in the summer, and in the Mediterranean during the winter. In the western Indian Ocean off Mozambique several schools were seen swimming north in March and around the Arabian coast they are most numerous during May.

In South-Africa they are very rare and the author only recorded them in November. The New Guinea observations were in the northern winter.

| immatures | Northern hemisphere calves are born in December and have about half the length of the mother. Cows with immatures are common round the Hebrides and Shetlands during the summer. |

In the Red Sea the author observed a school with small immatures in June, which would point to a southern hemisphere stock in the Indian Ocean.

| notes | In the Hebrides they are called Lowpers or Dunters. |

Off Manokwari in New Guinea 20 *Grampus* were on the 100 fathom line, which in that position is only $\frac{1}{2}$ miles from shore, standing on their heads in the water, tails in the air (lobtailing) in company with *Stenella malayana*.

There is only one record from the eastern Pacific Ocean, a specimen caught in Montery Bay California.

Three sea aquaria in Japan have up to half a dozen of these dolphins trained to give the same performance which the *Tursiops* demonstrate in Marineland of Florida and San Pedro.

A professional Bulgarian Dolphin hunter recognized this species from the draft illustrations for this guide, as abundant in the Black Sea, all the year round. The local name there is Avala and the measurements he gave was length 4 m and weight 600 kg.

The Australian records in the chart are not 100% certain. It has never been recorded in New Zealand.

genus	GLOBICEPHALA Lesson, 1828.
species	PILOT WHALE or BLACKFISH, plate nr. 65.
	(a) *G. melaena melaena* (Traill, 1809).
	(b) *G.m. edwardi* (A. Smith, 1834).
	(c) *G. scammoni* Cope, 1869.
	[= *G. sieboldi* Gray, 1846, probably the correct name].
	(d) *G. macrorhyncha* Gray, 1846.
area	See Chart 7.
	(a) temperate North Atlantic Ocean.
	(b) temperate Southern Oceans.
	(c) temperate North Pacific Oceans.
	(d) tropical Oceans.
sea temperature	Temperate regions 14 to 20° C, tropical 20 to 29° C.
length	Males to 28 feet (8.5 m), females to 20 feet (6.1 m).
weight	Males to 8400 lbs or 4 ton; 360 lbs/ft (3800 kg).
	Females ± 4000 lbs; 200 lbs/ft (1800 kg).
food	Squid and sometimes Fish (Herring).
teeth	4×10, in the front part of the jaws, diameter $\frac{1}{2}''$ (13 mm).
illustration	From life and photographs. It serves for all populations. The white marking on the belly (characteristic for (a) is grey to dark grey for the areas (b), (c) and missing in (d). It is invisible in the field because this dolphin does not jump out of the water, except when very young.
	The melon grows bigger and more bulbous with age, in females less than in males. Females and especially immatures have a head which resembles that of *Pseudorca* and *Grampus* (nr. 64). Individuals from (b) sometimes have a small white eyebrow and grayish saddle like *Orcinus* behind the dorsal fin.
special features	The conspicuous round head and heavy, rounded dorsal fin placed forward of the middle. The flippers are long and narrow. They cruise and surface very slowly, always in compact schools. The head shows first, producing a blast of 4 to 5 feet high, visible even in tropical temperatures, followed by the back and dorsal fin until about $\frac{3}{4}$ length of the animal is visible, then sliding back under water never showing the flukes.
	They are indifferent to ships, but do not flee and can be ap-

proached quite close. Once two immatures accompanied the author's ship alongside at 50 yards off, jumping just clear of the water in very flat arches. They were first taken for *Pseudorca* (66) until one spun round in mid-air and displayed the white mark on the chest. This was near Galita Islands (Tunesia).

speed
Cruising 2 to 4 knots, 14 knots easily and capable of 22 knots when excited.

breath
On the surface for several minutes regular breathing, then deep dive reaching depths of 3000 ft and can stay under water for 2 hours.

schools
5 to 20 normal, sometimes many hundreds (Far Ør June-Aug.).

biotope
Marine and coastal, outside the 100 fathom line and traveling along the edge of the continental shelf. Oceanic records are rare.

migration
There is undoubtedly a migratory movement between colder and warmer area's, but this is not clearly defined.
Population of (d) has shorter flippers and is perhaps "conspecific" with (b) which makes a separate tropical form uncertain.

immatures
These are difficult to record because of the wide range in size of the adults. Newborn calves of about 6 feet were observed in the Eastern Atlantic in August and November, in the Western Atlantic April and September, the Philippine Islands in May and off Panama in June. Immatures are seen all the year round.

notes
This genus strands quite often "normally" every second or third year, in large schools, which makes them rather well known because of the pictures in the newspapers.
All well-meant efforts to refloat and bring them back in deep water usually turn out to be futile, when released they swim straight back to the beached herd. This was, and perhaps, still is, a method to capture them in the Far Ør.
Some, including the leader are chased to the beach by small boats and the rest of the school follows suit.
It is a certainty that these animals in distress send out supersonic signals for help and that with dolphins the instinct to assist herdmembers in difficulties is stronger than that of

personal survival.

The schools do not associate with other genera, although in crowded areas other species might be near:
Delphinus delphis, Tursiops, Orcinus, Stenella euphrosyne or *Lagenorhynchus* in the North Atlantic; *Tursiops truncatus gilli, Stenella euphrosyne, Lagenorhynchus, Orcinus* or *Lissodelphis* in the N. Pac.

November 1965 a school of 30 Pilot Whales ascended the Thames river estuary as far as Tilbury Docks.

January 1967 a school of 40 stranded near Cape Kidnappers Hawkes Bay, New Zealand close to a large campsite. Because the odor of the dead whales would make life there impossible, about 40 men, without a boat, successfully refloated 36 of them. All seemed to be only semi-conscious. In about 4 to 5 feet of water, rolling, shaking and beating revived them and they swam back to open sea. The 4 largest individuals could not be moved.

genus	PSEUDORCA Reinhardt, 1862.
	FALSE KILLER (WHALE), plate nr. 66.
species	*Pseudorca crassidens* (Owen, 1846).
area	Tropical and subtropical oceans, occasionally temperate. Regularly observed in the northern Indian Ocean, rare elsewhere (Chart 1).
sea temperature	Above 20° C.
length	Males to 20 feet (6.1 m), females to 15 feet (4.6 m).
weight	Males to 4800 lbs, 240 lbs/ft (2170 kg).
	Females to 2400 lbs, 160 lbs/ft (1100 kg).
food	Squid and Fish (70 lbs/day).
teeth	4×8 to 11, conical $\frac{1}{2}$ to $\frac{3}{4}''$ diameter (12–18 mm).
illustration	Life and photographs. Uniform jet black, occasionally with starshaped scars. Sometimes a very dark grey marking on the underside similar to *Globicephala* and *Grampus* has been noted in the eastern Pacific. Immatures have grey undersides.
special features	The black colour, blunt head and a prominent cucumber-shaped dorsal fin 1 foot (30 cm) high. When approaching a ship they jump with low flat curves, just clearing the water.

They ride the bow wave of small boats, but are probably not fast enough to keep up with a big ship.

Those trying to intercept the ship were only seen riding and jumping the spreading bow wave far behind.

Lobtailing has been observed. Immatures seen in the Persian Gulf in February were playing and jumping together in the big school.

speed	Cruising 6 knots, maximum ± 12 knots.
breath	Every 12 to 15 seconds, immatures ev. 8 seconds.
schools	20 to 30, sometimes up to 50, subdivided into small groups of 4 to 6 animals, covering an area of $\frac{1}{2}$ mile diameter. Stranding of herds over 200 have been reported.
biotope	Marine and oceanic, occasionally coastal. Usually deep water but those in the Persian Gulf were in 20 to 26 fathoms (36–47 m).
immatures	Newborn calves are 5 to 6 ft (1.7 m), they were seen in February and March but probably are born all year round as in most schools all sizes are present.
notes	This species, like the Pilot Whale sometimes swims ashore in large groups, mostly in temperate climates (New Jersey, Great Britain, South Africa, Buenos Airos, Chatham Islands, Tasmania) but also in the tropics (Ceylon, Zanzibar). Some of these live till three days after the beaching and have been heard to squeal, which also has been noted at sea.

They are rare in the Atlantic and the Pacific, although in 1964 a large school was sighted off Hawaii Islands and two specimens were captured for the Sea Life Park in Oahu. From these no aggressive behaviour was reported.

Their "home" area seems to be the northern Arabian Sea where they are seen regularly and in numbers.

They were hunted here already in ancient times for the ivory of the teeth, which were traded across Asia, Alaska and Canada to the Indians of North America.

genus	PEPONOCEPHALA Nishiwaki and Norris, 1966. LITTLE KILLER (= BROAD BEAKED DOLPHIN and Many Toothed Blackfish), plate nr. 67.
species	*Peponocephala electra* (Gray, 1868).

120

	[= Lagenorhynchus electra (Gray, 1846)].
area	Tropical and subtropical oceans of the world (4 skulls from Solor – Indonesia; 1 from Madras – India; 1 from Car Nicobar; several from Western Australia; 1 from Oahu – Hawaii; 1 NW of St. Paul's Rocks – Atlantic Ocean). The author is convinced he observed these dolphins repeatedly between the Marquesas, Tuamotu and Society Islands in the Central Pacific Ocean (Chart 7).
sea temperature	Above 25° C.
length	8 to 9 feet (2.4–2.8 m).
weight	Probably ± 400 lbs (181 kg).
food	No records, probably Fish.
teeth	Upper jaw 24 to 25, lower jaw 2 × 24, small and pointed.
illustration	From life, photographs and description. The colour is very dark grey, becoming slightly lighter on the underside. No scars. Thin white lips and a patch of lighter colour similar to that in *Globicephala* between the flippers and round the genital slit.
special features	In outer appearence a small *Pseudorca* with a more pointed head. The dorsal fin, in the middle of the back has the shape of that of *Tursiops* and the size of that of the *Euphrosyne* dolphin. When swimming fast they only break surface with the blowhole, followed by the dorsal fin, the body stays under water. One school of 15 specimens was roaming round just under the surface, dorsal fins exposed and occasionally bobbed up and down, showing the head and upper body clearly with each blast, diving again with a strong arching of the last part of the tailstock which has a sharp ridge. They were not afraid of the ship, which passed the school at a distance of about 150 yards, but regularly stood upright in the water, head and neck out, looking round. They made no effort to ride the bow or stern wave.
speed	Cruising 2 to 3 knots, maximum estimated at 15 knots.
schools	The schools around the Central Pacific Islands are 5 to 15 individuals.
breath	At 10 to 20 seconds intervals.
biotope	Marine and oceanic.
immatures	The Oahu specimen was a newborn calf, found in August.

notes	Because the outer appearance is so much more a "Killer" type than any other group, the author used the name Little Killer instead of Broad Beaked Dolphin which was supposed to be a *Lagenorhynchus*.
	The skull of this animal was described in 1846, the first life specimen only as recently as August 1963 after it was found by a fisherman in a shallow bay on the South coast of Honshu (Japan).
	See also notes nr. 48: Illigan Dolphin.

genus	FERESA Gray, 1871.
	PYGMY KILLER (WHALE), plate nr. 68.
species	*Feresa attenuata* Gray, 1875.
area	Tropical and subtropical Atlantic and Pacific Oceans (Chart 7).
	Untill 1954 this dolphin was only known from 2 skulls of unknown origin. In that year one specimen was caught in southern Japan, but processed before it could be measured. In 1958 one was found near Dakar, then in 1963 fourteen were captured alive and studied in Japan.
	In 1965 one out of a school of 50 was caught in Hawaii and studied in the Sea Life Park on Oahu.
sea temperature	Probably above 15° C. The school of 14 near Japan, swam in 13½° C, but just outside the much warmer Kuro Shio. The author observed a school which he is certain were this species off Mossammedes (Angola) in water of 18° C.
length	Male 8 feet (2.4 m), female 7 feet (2.3 m).
weight	± 350 lbs; 44 lbs/ft (160 kg).
teeth	Upper jaw 10 to 11; lower jaw 11 to 13.
food	Sardines, it refused Squid and other Fish. 17 lbs/day (8 kg).
illustration	From sketch and photographs. There is a similar white area on the underside as in *Globicephala* (nr. 65) and *Grampus* (nr. 64).
special features	Extremely rare, similar to but smaller than the False Killer (nr. 66). The light grey colour on the flanks is sharply defined against the dorsal black. The fin is 10″ high. They show a habit of standing upright in the water, head out, looking round.

Not afraid of ships. Aggressive, snaps the jaws, beats flippers and flukes on the water and makes a growling sound as warning.

Another dolphin and a pilot whale in the tank with her became nervous; the pilot whale was dead next morning, killed by butting in the region of the throat.

It is also reported that dolphins will kill sharks by butting, when they show too much interest in their calves.

speed	Cruising 6 knots, certainly capable of more (over 20 knots).
breath	Every 6 to 7 seconds, deep dives of 25 seconds to 3 minutes.
schools	10 to 50.
biotope	Deep water, coastal but probably also marine and oceanic.
immatures	In January foetuses 8″ (20.5 cm) and 22″ (53 cm). Calves probably born in spring and summer. Newborn calf July 3 feet (91 cm).
notes	With Pygmy Right whale (nr. 89) and Little Killer (nr. 67) the rarest of whales. The first full description was given by Dr. Masaharu Nishiwaki.

The captured Japanese school was brought to the ITO Aquarium Honshu and died within 22 days from what was thought to have been pneumonia. Only one male took food. The school, off Mossammedes (July, 1966) was milling round in an area of about $\frac{1}{2}$ mile exposing the high dorsal fin more than other dolphins, not jumping but frequently beating with the flukes on the water. A big male Killer Whale (nr. 69) was observed 3 hours earlier further south.

genus	ORCINUS Fitzinger, 1860.
species	[= *Orca* Gray, 1860].
	KILLER WHALE or ORCA, plate nr. 69.
	Orcinus orca (Linnaeus, 1758).
	[= *O. rectipinna* (Cope, 1869)].
area	All oceans, only numerous in sub-arctic, subantarctic, and temperate regions indicated in the Chart 7.
sea temperature	0–30° C, only frequent below 10° occasionally 15° C.
length	Males: to 30 feet (9.2 m), females: to 20 feet (6.1 m).
weight	Males to 8 ton, 400 lbs/ft, a 24 feet male weighed 8000 lbs (3600 kg), females to 1$\frac{1}{2}$ ton, 150 lbs/ft.

food	Octopus and Squid, Dolphins, Gray Whales, Minke Whales, probably Right Whales, Seals and Fish including large rays, sharks and Basking Sharks.
teeth	4 × 10–12, strong, conical and interlocking, diameter 2″ (5 cm) and oblong, larger across the beak.
illustration	From life and photographs of a Northern hemisphere specimen. Most existing illustrations show the male's dorsal fin too small. It reaches a height of 6 feet. The underside of the flukes is white. Flippers reach a length of 20% of the body in old males. The Southern hemisphere Killers have only a thin white "eyebrow" and a more massive melon.
special features	The black and white colouring is conspicuous, which combined with the large size and high dorsal fin makes recognition not difficult. They always show a good deal of themselves when blowing. Their food indicates an aggressive nature, fearing nothing and nobody. Consequently they are indifferent to ships and can be approached quite close. Lobtailing has been observed and they sometimes stand vertically in the water, head out looking round, occasionally jumping right out at an angle of about 45°, falling back with a big splash. The thin white eyebrow of the southern population is difficult to detect in the field.
speed	Normally cruising at 4 knots, capable of much more and reported to be able to reach 30 knots.
breath	4 or 5 times at 10 to 20 seconds intervals, than a deep dive lasting $1\frac{1}{2}$ to 6 minutes.
schools	2 to 8, sometimes as many as 30 or 40. In all schools males are present. A school of 3 is usually one male with 2 females. Singles are always males.
biotope	Coastal to marine, rarely oceanic. Sometimes close inshore and in shallow water (8 in Auckland Harbour October 1968) but normally in deep water.
migration	They go to warmer waters in the spring and summer. On the Pacific coast of Middle America they are most often seen in May. The author observed 6 near Minicoy Island Indian Ocean in January. In New Zealand they appear off North Island around the beginning of November, accompanied by newborn calves. Small schools often spend about 3 weeks

in Hawkes Bay, daily making the same circuit along the coast and reefs, before going further north (information Mr. F. Robson, Taradale Napier). During December and January several schools were sighted between North Island, New Zealand, Tongga Islands and Tahiti. They return to colder areas during March and April.

immatures   Calves are born in late autumn or winter (Nov.-Dec.; June-July) have a length of 6–7 feet. Immatures have been reported to ride the bow wave of small boats.

In very young animals the white is yellowish.

notes   A strong, ruthless and greedy predator of all life in the Ocean, who will attack, thrash and tear to pieces practically every imaginable prey. They break up ice 3 feet thick to reach seals and walrus they have seen resting there, looking over the top.

The photographer of one of Scott's Antarctic expeditions once barely escaped the same fate.

Hunting in "wolfpacks" they "play" with their victims in a surrounded school of seals or dolphins up to 20 minutes, throwing them up in the air and bumping them out of the water from underneath, during the pursuit.

There have been no reports of attacks on boats or people. Perhaps boats are to them as the cars in game reserves, which wild animals do not associate with food or danger as long as the occupants stay inside.

In one stomach the remnants of 13 dolphins and 14 seals were found, in another that of 32 seals, according to Slijper. One large male nearly caught a dog, barking at it from a low rocky shore, beaching himself partly for the purpose. A little dog in the Sea World Aquarium of San Diego did not escape.

A young male was captured alive once near Vancouver Island, towed into Vancouver Harbour and kept alive there in spite of one harpoon in his body and several bullet wounds. It died after 3 months and did not take food the first 55 days.

The cause of death was very likely the polluted water of the harbour.

A 3,5 ton male was later caught in the same region and

towed in a surrounding net 450 miles to Seattle USA for the Seattle Public Aquarium. It became quite tame and very friendly and cooperative with its owner Edward J. Griffin, who wrote an excellent article about his experiences in the National Geographic Magazine March 1966. It drowned in the nets of the enclosure in the autumn of that year, trying to escape when a school of other Killers passed.

Since then many marine establishments acquired Killers, who have great show value and prove to be very intelligent and easy to train.

The antarctic whaling fleets encounter schools of 10 to 20 daily. They attack and devour the freshly killed whales and are especially fond of the tongue. This blown up tongue gives the dead whale its buoyancy. Since the catcher crews receive no bonus for a lost whale they fight and shoot the Killers. As soon as one is wounded and bleeding the rest of the pack attack and eat it creating a diversion which usually enables the boat to bring in the whale safely.

Killers occasionally strand, mass strandings are rare, one is recorded from Vancouver Island and another, involving 15 animals, 13th May 1955 on a low sloping beach in the NW entrance of Cook Strait near Wellington, New Zealand.

ALULA WHALE or ALULA KILLER, plate nr. 70.

This large dolphin was seen by the author on several occasions in the eastern Gulf of Aden, north of the village Alula, west of Cape Guardafui which is the extreme NE point of Africa.

Its appearance is quite different from any of the known species and it has therefore been given a separate paragraph.

area	Eastern Gulf of Aden (Chart 7) to Socotra.
sea temperature	26 to 30° C.
length	Estimated 20 to 24 feet (6–7 m).
weight	Estimated about 4000 lbs (1800 kg).
food	No records.
teeth	No records.
illustration	From life. The author and several other officers standing on

the bridge made sketches of the animal as it was swimming past in a flat calm sea. From these the painting was made. The colour was sepia brown and showed white starlike scars on the body. The dorsal fin, well above the surface was very prominent.

special features A rounded forehead, similar but not quite as round as in *Globicephala*. A little snout. The dorsal fin was estimated to be at least 2 feet high.

At first encounter a school of 4 approached the ship head on and seeing the dorsal fins the author thought they were *Orcinus orca* (nr. 69). When they passed the ship at a distance of less than 50 yards just under the surface in the flat calm, clear sea, it was obvious that this was a different species. They were indifferent to the ship and neither changed course nor dived.

speed Cruising ± 4 knots.
breath Every 10 to 20 seconds.
schools 4 to 8, usually 6.
biotope Deep coastal waters in Gulf of Aden.
immatures Not observed.
notes These dolphins were seen in the area during crossings in April, May, June and September, usually swimming just under the surface with the dorsal fin above the water. One duty officer reported he observed them chasing a school of smaller dolphins, who tried to escape. There is, however, a possibility that both species were chasing the same prey.

Attack and devouring of live whales by predators in this area was witnessed by at least two colleagues of the author, who could not identify the species, however. One of the victims was a Sperm Whale.

GROUP XV   IRRAWADDI DOLPHIN
Genus *Orcaella* Gray, 1866.
[= *Orcella* Anderson, 1871].

As the latin name indicates, the *Orcaella* is related to the *Orcinus orca*. It looks like a dark version of the Beluga or White Whale (nr. 5), and with *Sousa* and *Sotalia* (nrs. 53-59) share an estuary and river life with similar habits.

	IRRAWADDI DOLPHIN, plate nr. 71.
species	*Orcaella brevirostris* Gray, 1866.
area	Coasts and estuaries of South East Asia, perhaps to Northern Australia (Chart 8).
subspecies	*Orcaella brevirostris fluminalis* Anderson, 1871.
area	From 300 to 900 miles up the Irrawaddi River (Birma); also seen 70 miles up Pussur River (East Pakistan).
sea temperature	25 to 30° C.
length	± 7 feet (2.1 m).
weight	± 220 lbs (100 kg), the freshwater subspecies is probably lighter.
food	Fish.
teeth	2 × 12 in the upper jaw, 2 × 19 in the lower jaw. Small and conical, diam. 0.2″ (3 mm), wearing off considerably with age.
illustration	Life and other illustrations. The colour is very dark grey. The subspecies is a lighter grey and has white undersides.
special features	When this dolphin surfaces the round head is conspicuous, followed by the back, dorsal ridge and little fin. It comes up in a leisurely way, then dives in a hurry with strong arching of the tailstock. The blowhole is situated left of the middle.
speed	2 to 4 knots, enough to make headway in the strong currents they seem to prefer.
breath	2, 3 or 5 times every 10 seconds, then normally a deep dive of 20 to 70 seconds, occasionally to 3 minutes.
schools	3 to 6, usually 4.
biotope	Muddy river entrances with strong tidal currents, often brackish water. They swim against the tide.

immatures	In September calves of $\frac{1}{2}$ length were observed in Malaya and Borneo. Newborn calves are 0,4 length of the mother.
notes	The author has seen them only once jump clear of the water, this was during October in the Klang River and they were quite numerous at the time.

The grey colour, quiet ways and the muddy habitat require a sharp lookout to detect them.

The local name in Borneo is Lumbalumba. They are regularly seen in: Deli River (Sumatra), Klang River (Malaya), Rajang River (N. Borneo), Surabaya and Tjilatjap (Indonesia) and less frequent in Penang (Malaya), Makassar (Celebes) and Southern New Guinea (Indonesia); skulls have been found in Northern Australia.

This family is not divided in subfamilies as the *Delphinidae* but can be split in three groups. The most important feature, as the name implies is a prominent beak. In most species the lower jaw is slightly longer than the upper jaw and has one or two pairs of teeth against none in the upper jaw.

On the throat is a V-shaped groove, pointing forward (rudimentary expansion groove?). The dorsal fin is small, triangular and placed well back of the middle. The pectoral fins are very small for the size of the animals and are placed far forward.

Their slow and quiet ways when coming up to blow, make them difficult to detect. From what the author observed there is good reason to believe they are shy and spend most of their time well below the surface.

When a ship approaches, they sound instead of swimming away and if high in the water occasionally show the flukes.

GROUP I   Genus *Tasmacetus,* one species, extremely rare and only partly known from 6 stranded and damaged specimens in S. New Zealand.

GROUP II   This group have long cigar shaped bodies, tapering at both ends, small heads with rather long pointed beaks, which gradually, without much demarcation, go over in the forehead. The flukes are wide, 25 % of the body's length and have no median notch. There are 2 genera:

Genus *Mesoplodon* 12 species
Genus *Ziphius*   1 species

The *Mesoplodon* are, unofficially, subdivided in 6 *Mesoplodon* with one pair of small teeth which in addition seem to have round tips on the flukes; and 6 *Diplodon* with one pair of large teeth. The tips of the flukes seems to be more pointed.

Two *Mesoplodon* species are only known from few skulls each, which have been subject to classification and reclassification by different authorities. The question whether they belong to either one or the other genus of Beaked Whales was not definitely solved until 1968.

All species are rare to very rare and only known from a few dead and stranded animals. Only in Japan they have been accidentally caught alive on Tuna fishing lines.

The *Ziphius* is less rare and has regularly been observed by the author in all oceans, always in very deep water, over 2000 fathoms, giving the impression that that is where they live and die, sinking to great depths, probably the reason why so few are washed ashore.

Beaked whales which the author believes to have been *Mesoplodon* were seen in similar schools as *Ziphius* and also in very deep water: Philippines trough, Japan trough, the Mexican Deep in the Pacific, the Somali and Arabian Deeps in the Indian Ocean and the Congo Deep in the South Atlantic.

In the North Atl. most records are west of the meridian of the Azores.

GROUP III   This group is better known. They are torpedo-shaped with a cylindrical body and a round head, but tapering towards the tail. A long round beak is clearly demarcated from the forehead.

The large flukes have a spread of 27 to 31 % of the body's length.

They are:

Genus *Berardius*   2 species
Genus *Hyperoodon* 2 species

Both genera belong to colder climates and have a representative in the northern and southern hemisphere. The northern species have been hunted already more than a century for their oil.

The southern population escaped this fate and are therefore less well known. The quoted weights in this chapter are all estimates, for all other whales and dolphins an indication could be found somewhere.

The type of animal was taken into consideration and curves used, constructed from the little and scattered information available.

Most scientific publications mention no weights except for some specific organ and little else can be done. To ascertain the weight of one of these stranded whales somewhere on a deserted beach is impossible and the few small local whaling factories in Japan, where *Berardius* is sometimes caught and processed, are not interested except in the amount of oil it yields. This is 2 tons for an animal of 36 feet.

genus	TASMACETUS (Oliver, 1937).
	TASMAN (BEAKED) WHALE, plate nr. 72.
species	*Tasmacetus shepherdi* Oliver, 1937.
area	A total of 6 specimens washed up on beaches of Stewart Island, Banks Peninsula and Cook's Strait, east coast New Zealand (Chart 8).
sea temperature	Probably temperate or subantarctic.

131

length	Males 20 to 30 feet (6.1–9.1 m), the only known female was 18 feet (5.5 m).
weight	Estimated at 5400 lbs (2450 kg) for a 20 ft specimen.
food	No records, but probably Squid and Fish.
teeth	Upper jaw 2 × 19, lower jaw 2 × 26, height: 13–27 mm, width 15 × 5 mm, two large teeth in the tip, height 42 mm, width 24 × 16 mm.
illustration	Made from the original description, available measurements and a photograph.
special features	This whale looks like a large Beaked Whale, intermediate between a Bottle-nosed Whale (nr. 84) and *Mesoplodon*. The dorsal colour is dark brown becoming lighter on the flanks with white underside.
speed	No records.
breath	No records.
schools	The strandings were once two males, the others were single.
biotope	No records. Because it is so rarely observed, the species is probably subantarctic and oceanic.
immatures	The female carried an unspecified foetus (in spring).
notes	The numerous teeth in both jaws besides the 2 larger ones in the tip of the lower jaw, made the creation of a separate genus and species necessary. Outward appearance and the 2 large teeth bring it in the range of the Beaked Whales.

It is here interesting to note that the young of *Ziphius cavirostris* (nr. 81) also have 28 to 30 pair of teeth in the gums which do not develop.

The female stranded in spring near Wanganui, the others during the summer. Maximum circumference almost equals total length.

genus	MESOPLODON Gervais, 1850.
	Sowerby's (Beaked) Whale, plate nr. 73.
species	*Mesoplodon bidens* Sowerby, 1804.
area	Gulfstream area of the North Atlantic Ocean (Chart 8).
sea temperature	10 to 20° C.
length	16 feet (4.9 m).
weight	± 3000 lbs; 180 lbs/ft (± 1460 kg).
food	Mainly Squid, sometimes Fish.

| teeth | Males one pair of small or moderate triangular teeth situated about $\frac{1}{3}$ length from tip of the lower jaw, just at the mandibular symphysis (where right and left jaws join). Height $3\frac{3}{4}''$, length at base $1\frac{1}{4}''$ (93 $\times$ 31 mm). Females have a very small pair of teeth in the same position or none at all. The long axis has a strong ($\pm$ 45°) inclination backwards, the denticle (a sharp hard point at the apex of the tooth) is situated on the anterior point of the apex. |

Place of the teeth is indicated by the anterior margin and expressed in fractions or % of mandible length.

illustration	From different available illustrations and description.
special features	The body is covered with white stripes. These are healed scars of wounds probably inflicted during fights amongst each other. The flukes have no median notch.
speed	No records.
breath	No records.
schools	No records.
biotope	Probably oceanic and marine in deep water.
immatures	Calves are born in spring. Length 6 feet (1.8 m).
notes	The chart indicates where strandings were reported. This whale has never been reported from observations at sea. One stranded specimen survived three days and was reported to make a noise like a cow. One specimen was found on Sicily.

GULFSTREAM (BEAKED) WHALE, plate nr. 74.

species	*Mesoplodon europeus* Gervais, 1855. [= *M. gervaisi* (Deslongchamps, 1866)].
area	The Gulfstream from Trinidad to the English Channel (Chart 8). The range is just southward of that of the True (Beaked) Whale (nr. 75).
sea temperature	15 to 30° C.
length	14 to 22 feet (4.3–6.7 m).
weight	$\pm$ 2000 to 6000 lbs, 150–270 lbs/ft (900–2700 kg).
food	Probably Squid.
teeth	Males one pair, triangular, moderate size (96 $\times$ 15 $\times$ 14), laterally compressed, teeth situated about $\frac{1}{7}$ length from the tip of the lower jaw, placed before the mandibular symphysis (where the right and left jaw join). The long axis has

a slight inclination forward. Females have no teeth.

illustration   From available illustrations and photographs of the True
                    (Beaked) Whale, which is a practically identical species.
special features   The body is covered with white stripes. These are healed
                    scars of wounds probably caused by fighting amongst each
                    other. The flukes have no median notch and round tips.
speed   No records.
breath   No records.
schools   No records.
biotope   Probably oceanic and marine, deep water.
immatures   A mother with a newborn calf were found in Jamaica on
                    21 February 1953. The lengths were respectively 14 feet and
                    7 feet.
notes   The chart indicates the places of known strandings. The
                    species is only known from 12 animals. The only outward
                    difference between this Whale and the Sowerby's Whale
                    (nr. 73), are proportionately smaller flippers.

TRUE'S (BEAKED) WHALE, plate nr. 75.

species   *Mesoplodon mirus* True, 1931.
area   Temperate North Atlantic Ocean, range seems to be just
                    north of the Gulfstream (Beaked) Whale, it has also been
                    reported from the temperate South Atlantic, a stranding east
                    of Cape Agulhas, Sth. Africa (Chart 8).
sea temperature   10 to 20° C, perhaps 25° C.
length   17 feet (5.2 m).
weight   ± 3000 lbs, 180 lbs/ft (1400 kg).
food   Squid and sometimes Fish.
teeth   Males one pair, small or moderate size ($1\frac{3}{4} \times \frac{3}{4} \times \frac{3}{8}''$) (49
                    × 17 × 9 mm) triangular sideways compressed teeth in the
                    tip of the lower jaw. They only erupt in older males and not
                    in females. The long axis has a slight inclination forward.
illustration   From photographs and description.
special features   No mention is made of scars and in the available photo-
                    graphs no visible scars can be seen.
                    The flukes have no median notch and the tips are round.
speed   Cruising at 3 knots, surfacing at an angle of 10–20°, when
                    sounding the angle increase to 30°, and speed to ± 6 knots.
breath   On the surface between deep dives with 5–10 sec. intervals,

	swimming only a few feet below the surface.
schools	Probably single or in pairs. These data are from an observation by the author of a *Mesoplodon* in the N. Atlantic Ocean (95′ NW of Chaucer Bank) which he believes to belong to this species. The colours were light brown (as in fallow deer) dorsally, lighter on the flanks to very light, but not white below. There were no visible scars.
biotope	Oceanic in the northern and colder half of the Gulfstream, deep water.
immatures	A newborn calf of 7′2″ was found with a mother of 17 feet in March.
notes	A total of 14 stranded animals is known. They have not been observed with certainty at sea. The chart indicates the areas where strandings took place.

The difference in mean latitude between this and the previous species is from the SW Coast of Ireland to the English Channel on the east side of the Atlantic Ocean and between Nova Scotia and Long Island on the west side.

In 1960 a whale of this species was identified stranded on the south coast of South Africa.

The photographs and description which enabled the author to make the illustration are published in a paper presented to him by Dr. Joseph Curtiss Moore.

	SCAMPERDOWN WHALE, plate nr. 76.
species	*Mesoplodon grayi* Von Haast, 1876.
area	Southern subantarctic and temperate oceans, once found on the Netherlands coast (Chart 8).
sea temperature	5 to 20° C.
length	12 to 13 feet (3.6–4.0 m).
weight	± 1700 lbs, 140 lbs/ft (770 kg).
food	Not recorded.
teeth	Males: one pair onion shaped small to moderate sized (± 70 × 80 mm) teeth about 33 % from the tip of the lower jaw, before the mandibular symphysis. In females the teeth do not erupt. The long axis has a slight inclination forward. In each upper jaw from the level of the single tooth backwards, a 10 cm row of 17 to 22 small but functional teeth (5 × 2 to 10 × 3 mm).

illustration	Life and description. The colour is very dark gray. The beak as far as the forehead and the chin are white, the face is often covered with numerous white spots.
special features	The flukes have no median notch, and the tips are probably round. The body is covered with white stripes, healed scars from fighting and also with large oval shaped spots, similar to those in other beaked whales.
	The author observed a school of 6 of these animals 400 miles east of North Island NZ, who were jumping out of the water about $\frac{1}{2}$ mile ahead of his ship.
	They shot up at an angle of about $30°$ with the surface as far as $\frac{2}{3}$ length, then fell back with a big splash. Swimming in "single file", they kept this up for about 5 minutes, then dived for the approaching ship and were not seen any more.
speed	3 to 4 knots.
breath	On the surface every 10 to 30 sec.
schools	2 to 6.
biotope	Marine and oceanic, occasionally coastal.
immatures	Calves are born in spring and have a length of 7 feet (2.1 m).
notes	The distribution is similar to that of the Straptoothed Whale (nr. 79). The author perhaps met another school of these whales (or nr. 79) swimming just under the surface, on the route from Sydney to Wellington, in much the same way as Cuvier's Whales (nr. 81).
	There are 20 known strandings between 1870 and 1968 in New Zealand, along the eastern coasts from North Cape to Stewart Island and inside Cook Str.; 5 from the Chatham Islands, one from Sth. Africa. The strandings are indicated in the chart.
	In February 1968, two beaked whales, probably of this species, were seen off the breakwater of Napier in Hawkes Bay (NZ). One remained there for 3 hours before disappearing again to sea.

The next 2 species of *Mesoplodon* received no number in this guide. They were only known from respectively one and three skulls, and identification was uncertain. During 1968 both species became valid by discovery of additional skulls.
No illustration, however, can be made.

LONGMAN'S BEAKED WHALE
Mesoplodon pacificus Longman, 1926 or
Indopacetus pacificus Longman, 1926.

It was described from a skull found near Mackay, Queensland, Australia. In March 1968 Dr. Maria Louisa Azzaroli described a second skull found in 1955 near Mogadiscio (Somali) which establishes the identity of a separate species.
The live animal is larger than other members of the genus; judging from the skulls, over 23 feet (7.0 m) which is in the range of *Ziphius*.
Dr. J. C. Moore the world's expert on Beaked Whales has placed it in a new Genus: *Indopacetus*.
There is a possibility the author observed these animals in the Gulf of Aden and the Sokotra area, being very large Beaked Whales, and certainly not *Ziphius*. In that case the colour is light-brown to rust-brown with a lighter head and slender beak. Shape similar to *Ziphius*. The teeth are placed in the tip of the mandible and have a forward inclination.
Mr. K. C. Balcomb of Pacific Beach, Washington USA took a photograph of a school of 25 Beaked Whales on the equator at 165° West, which were almost certainly this species.

Mesoplodon hectori Gray, 1871.
HECTOR'S BEAKED WHALE

Originally described from three skulls of very immature, perhaps neonatal calves, found in New Zealand waters, as *Berardius arnouxi*. Since then they were reclassed several times untill Dr. J. C. Moore in 1968, definitely established them as a valid *Mesoplodon* species. In 1967 the skull of an adult female was found in Tasmania, which settled the controversy.
The teeth are placed 4,5 % behind the tip of the mandible and have a forward inclination. Nothing of the outward appearance or habits is known.

SABRE-TOOTHED WHALES

Until 1963 only one Sabre-Toothed Whale was known (as *Mesoplodon stejnegeri,* True 1885). In that year Dr. Joseph Curtiss Moore divided the species in three valid species by the shape and place of the teeth. Identification is only possible on close inspection of a stranded animal or the skull.

SABRE-TOOTHED WHALE, plate nr. 77.

species	*Mesoplodon* (*Diplodon*) *stejnegeri* True, 1885.
area	North Pacific Ocean from Oregon to Japan and Bering Sea (Chart 8).
sea temperature	5 to 15° C.
length	± 16 ft (4.9 m).
weight	± 2800 lbs, 175 lbs/ft (1270 kg).
food	Squid and perhaps Fish.
teeth	One pair in the lower jaw, approx. 28 % behind the tip. The anterior margin behind the mandibular symphysis. Shaped roughly triangular. In males much bigger than in females.

H: $5\frac{3}{4}''$ l: $8\frac{1}{4}''$ w: $3\frac{1}{4}''$ (140 × 200 × 80 mm) (Moore: 160 × 95 × 15 mm) and slightly curved round the rostrum (f: 60 × 70 × 100 mm).
The long axis has an inclination backward and the denticle is situated on the anterior point of the apex.

illustration	From photographs and description.
special features	The flukes have no median notch. The tips are pointed. The body is covered with white stripes, the healed scars from fighting amongst each other, also with round white spots with a diameter of 2 to 3" (5–7$\frac{1}{2}$ cm) from an unknown cause.
speed	Probably 3 to 4 knots with a maximum of 6 knots.
breath	Not recorded.
schools	Single to schools of 2 and 3.
biotope	Probably Oceanic and Marine, deep water. Subarctic.
immatures	Calves are probably born in spring.
notes	Known from 10 strandings in the N. Pacific Ocean. The photographs and particulars which made the illustration of this rare whale possible, are from 2 articles by Robert T. Orr and 2 publications by J. C. Moore.

HUBB'S WHALE.

species	*Mesoplodon* (*Diplodon*) *carlhubbsi* Moore, 1963.
area	North Pacific Ocean, south of the Columbia River (west coast) USA and south of the north coast of Honshu in Japan (Chart 8).
sea temperature	10 to 20 perhaps 25° C.

teeth	Similar to 77. The anterior margin of the teeth is slightly in front of the mandibular symphysis, they are placed vertically and are straight, not curved, 20 % behind tip mandible. Size male: 160 × 90 × 15 mm. The denticle is situated a little forward of the middle of the apex.
notes	Known from 5 strandings, 1 in Japan, 4 between Washington and Lower California. In 1969 a skull was found on White Island-Bay of Plenty, northcoast of North Island New Zealand, which was identified by Moore as belonging to this species. The author on several occasions observed *Mesoplodon* Whales in oceanic positions in western and eastern Pacific Ocean, south of the tropic of Cancer. These perhaps could belong to this species. The author discovered a tooth in the Auckland war memorial Museum of an unidentified. *Mesoplodon* stranded Sept. 1946 which perhaps belongs to this species.

ANDREW'S WHALE

species	*Mesoplodon* (*Diplodon*) *bowdoini* Andrews, 1908.
area	Southern oceans, known from strandings in South Island New Zealand and Australia (Chart 8).
sea temperature	Below 15° C.
teeth	Similar to 77 but more triangular. The anterior margin of the tooth is slightly behind the mandibular symphysis and they are placed in a direction about 20° outward of the vertical. Size male: 138 × 75 × 14 mm, 22 % behind tip mandible.
notes	Known from 6 strandings in New Zealand between Cook Str. and Stewart Island, and one from Western Australia.

GINKGO or JAPANESE WHALE, plate nr. 78.

species	*Mesoplodon* (*Diplodon*) *ginkgodens* Nishiwaki and Kamiya, 1958.
area	Japanese waters, south to Taiwan-Formosa (Chart 8).
sea temperature	10° C.
length	17 feet (5.2 m).
weight	± 3200 lbs, 190 lbs/ft (1450 kg).
food	Probably Salmon.

teeth	One pair of large "onion-shaped" teeth in the lower jaw, situated on a bony crest $\pm$ 10″ (250 mm) behind the tip and showing outside the mouth. The teeth are sideways compressed, bigger in males than in females, height $2\frac{1}{2}$″, length 4″ (65 $\times$ 115 mm) 29 % from tip mandible, behind symphysis.
	The long axis has a slight inclination forward.
illustration	From photographs of a similar beaked whale and descriptions. The colour of immature female (7 feet, 2.4 m) was dark grey above and whitish underneath including the underside of the flukes and flippers. An adult was dark grey and turned jet black 2 days after death.
special features	The flukes have no median notch, they are even a little convex in the middle. The spread is 30% of the body's length, the ends are pointed but not sharp. The body is covered with many white stripes which are healed scars. Many are parallel and spaced 3″ (8 cm) apart, which indicate that they are from wounds which were caused by fighting of males. One specimen had one round scar ($1\frac{1}{4}$″ = 3 cm) and another many more of these.
speed	Cruising 3 to 4 knots. Because they catch fish a greater speed will be possible.
breath	Not recorded.
schools	The three animals caught were all solitary.
biotope	Coastal and marine, also in shallow water.
immatures	Born in spring, length 7 feet (2.1 m).
notes	Known from 3 specimens which were caught on fishing lines in the Japan Sea during spring 1963. In the field they can only be distinguished by the shape of the teeth. The name refers to the shape of the male's teeth, which is similar to that of the leaf of the Ginkgo tree.
	The photographs which enabled the author to make the illustration were put at his disposal by Dr. Masaharu Nishiwaki formerly of the Whale Research Institute of Tokio in Japan.

STRAPTOOTHED WHALE, plate nr. 79.

species	*Mesoplodon* (*Diplodon*) *layardi* (Gray, 1865).
area	Southern oceans, subantarctic and temperate (Chart 8).

sea temperature	5 to 15° C.
length	16 feet (4.9 m). The size of some of the skulls in the Dominion Museum, Wellington, suggests that some specimens reach lengths of over 20 feet.
weight	± 2800 lbs, 175 lbs/ft (1270 kg).
food	Not recorded.
teeth	One pair of strap-shaped teeth in the lower jaw. The posterior margin of the teeth level with the mandibular symphysis. The anterior margin about 26 % from the tip. They are long, narrow and flat in males (almost resembling a piece of rib), much smaller in females. Respectively 130 (and longer) $\times$ 36 $\times$ 15 and 40 $\times$ 26 $\times$ 6 mm, convex on the outside, slightly tapering towards the tip and tilted backwards at an angle of ± 45°. The denticle (sharp hard point at the apex of the tooth) in this species is bent outwards almost at right angles with the vertical. In young animals and females this is probably the only part which shows outside the gum.
illustration	From photographs of stranded animals. Judging from live sightings of this species or nr. 76 the colour is perhaps lighter grey.
special features	The flukes have no median notch and the tips are pointed. The body is covered with white scars of both the stripe and round variety as in most species of this genus in the Pacific Ocean.
speed	Probably 3 to about 6 knots.
breath	Not recorded, but during shallow dives probably at 10 to 30 second intervals.
schools	Not definitely recorded but probably 3 to 5 animals.
immatures	One female stranded in New Zealand during September and had obviously just given birth to a calf, indicating that parturition takes place in early spring. Length probably 7 feet (2.1 m). Two females stranded in late autumn and early winter had a small foetus each.
biotope	Oceanic and marine, deep water. Occasionally coastal.
migration	From the strandings no migration can be deduced. Females strand more often than males and mostly during autumn and winter.
notes	Known from ± 37 specimens only, all in the southern oce-

ans. In New Zealand 16 on the eastcoast of South Island between Otago and Cook Strait; 2 in Hawkes Bay; 1 in Hauraki Gulf and 1 in the Chatham Islands. From South America 1 is known, from South Africa 1, and from South and West Australia 6.

The photographs which enabled the author to make the illustration were put at his disposal by Dr. Falla of the Dominion Museum, Wellington (NZ).

BLAINVILLE'S (BEAKED) WHALE, plate nr. 80.

species	*Mesoplodon* (*Diplodon*) *densirostris* de Blainville, 1817.
area	All tropical and temperate oceans (Chart 9).
sea temperature	15 to 31° C.
length	15 feet (4.6 m) perhaps longer.
weight	± 2400 lbs, 160 lbs/ft (1100 kg).
food	Not recorded.
teeth	Males one pair of triangular teeth with a very long root in the lower jaw, 35 % behind the tip and posterior to mandibular symphysis (joint of right and left lower jaws). The crown is smaller than in the Sabretoothed Whale, but the total length is greater and the lower jaw forms a big crest to accommodate the root. Height 6″, length 3¾″ (150 × 94 × 44 mm). Females have no visible teeth, (54 × 30 × 7 mm).
illustrations	From description and photographs.
special features	The body is covered with white scars, long stripes probably formed from the healed wounds of fights amongst each other. The tail has no median notch.
speed	Cruising 2 to 3 knots.
breath	4 to 5 times every 6 to 10 seconds, then a deep dive.
schools	3 to 4 occasionally 6.
biotope	Oceanic and marine. Deep water.
immatures	A calf was seen in April.
notes	Known from 14 stranded and two live specimens. The coast-records indicate in the chart where stranded animals have been found. The open ocean records are, where whales believed to belong to this genus and more particular to this species have been observed by the author. It is the only member of this genus which is known to live definitely in the

tropical zone. The fact, however, that all other species stranded in colder areas does not exclude the possibility that they also live in the tropics. There are only two records from the Pac. ocean, one from Norfolk Island and one from New Guinea.

genus	ZIPHIUS G. Cuvier, 1823.
	CUVIER'S WHALE or GOOSEBEAKED WHALE, plate nr. 81.
species	*Ziphius cavirostris* G. Cuvier, 1823.
area	All tropical oceans, occasionally to temperate latitudes in the summer (Chart 9).
sea temperature	15 to 30° C.
length	Males 20 to 22 feet (6.1–6.7 m); females 20–23 feet (6.1–7 m). Females are approximately 1 foot longer than males. Lengths of 28 feet (8.5 m) have also been reported.
weight	5700 to 11000 lbs, 260-400 lbs/ft (2.6-5 tons).
food	Fish and probably Squid.
teeth	Males: one pair big blunt teeth in the tip of the lower jaw, which extends beyond the tip of the upper jaw. In females the teeth do not erupt. Immatures were caught which had 28 to 30 tiny sharply pointed teeth in each jaw under the gum.
illustration	From life and photographs. The majority seem to be dark in different shades of black and grey. In nearly all schools which the author observed there was at least one partly creamy white above and dark grey, chocolate or rusty brown over the rest of the body. The brown colours occur mostly in the Indian Ocean and these were chosen for the illustration.
	The dorsal fin in specimens from the Pacific and Atlantic Oceans appears to be higher than in the Indian Ocean.
special features	They often swim slowly and lazily just under the surface or dive to about 6 feet. When blowing show the round forehead, the back and dorsal fin, usually not the beak. The flukes never break the surface. Because they are indifferent to ships and do not take avoiding action until the ship is about 500 yards off, sometimes less, it is easy to observe them. When sounding some use a steep dive, others just

"sink" out of sight. Jumping was noticed on two occasions. The bodies are usually covered with white scars, males, females and immatures alike, caused by the teeth of aggressive males.

In the Pacific population the round white spots are also sometimes present. The flukes have a slight median notch.

speed   Cruising 2-3 knots, 10 knots when escaping, and as once happened, trying to keep pace with a ship.

breath   2 or 3 times at 10 or 20 seconds intervals, then sounding for one minute to half an hour.

schools Normally 4 to 6, sometimes only 2 and occasionally 15.

biotope Marine and oceanic, always in deep water (2000 fathoms).

immatures Calves are born all the year round. Different females in Japan were carrying foetuses in August of 1, 3, 5 and 7 feet.

notes   Between 1948 and 1952: 85 were caught on the south and east coast of Honshu (Japan), 60% of these were males. In the N.E. Pacific only 5 strandings were reported in 13 years. There is undoubtedly a migration from warmer to colder latitudes in the summer. In Japan they are only caught between May and September coinciding with seawater temperatures of 15 to 20° C.

The "creamy white" of older animals is an accumulation of many small white spots, which can still be distinguished separately where white and dark pigmentation meet on the flanks.

The 19 known strandings in New Zealand between 1870 and 1968 all occurred on the east coast north of Otago, 6 were in Cook Street and 2 in Chatham Islands. There is only one (unrecorded) stranding in the warmer waters of the north coast, a female with a newborn calf in Tauranga Harbour. Both were towed out alive to open sea, the mother washed up dead a few days later on the beach.

Allthough Beaked Whales are rare in the Mediterranian, Goosebeaked Whales regularly strand on the coast of the Ligurian Sea (Genoa).

genus	BERARDIUS Duvernoy, 1851.
	Arnoux Whale, plate nr. 82.
species	*Berardius arnouxi* Duvernoy, 1851.
area	Antarctic and subarctic waters near New Zealand and eastern South America (Chart 9).
sea temperature	0 to 10° C, occasionally 15° C.
length	32 feet (9.8 m).
weight	To 8.5 tons (600 lbs/ft).
food	Probably Fish and Squid.
teeth	Two pairs in the lower jaw. One pair of 3″ (7,5 mm) in the tip, which extends 4″ (10 cm) beyond the tip of the upper jaw, a second pair 8″ (20 cm) further back of 2″ (50 mm). The teeth do not erupt until sexual maturity, the second pair is sometimes missing. The shape is conical, slightly sideways compressed and they grow perhaps with a layer of cement each year abrading during life with continual use. Both males and females use them for fighting.
illustration	From photographs. The colour is black or dark grey above. The numerous white scars make the general effect lighter. The flanks and underside are lighter.
special features	The long cylindrical body with the round head and clearly demarcated snout. Spread of the flukes is 8 feet (2.4 m) which is 27% of the body's length. They are full of scars through fighting in the mating season. The head is 13,5–20 % of total length.
speed	Cruising probably 2 to 3 knots.
breath	Not recorded, but probably several minutes regular breathing on the surface, followed by a deep dive.
schools	Not recorded, probably small.
biotope	Coastal and marine in deep water.
migration	There is evidence that they migrate during the summer to warmer waters for calving and mating.
immatures	Calves are born in January. Length 9 to 14 feet (2.8–4.2 m).
notes	This species and the next are practically identical; Arnoux' Whale perhaps has smaller flukes, smaller flippers and a smaller head. It is only known from a few stranded specimens. The photographs taken by Mc.Cann which enabled the author to make the illustration, he received from Dr. Falla

of the Dominion Museum in Wellington, New Zealand. There are 15 known strandings in New Zealand (1870-1968) between Stewart Islands and Cook Street, two in Hawkes Bay, one in Hauraki Gulf. One is known from Australia, one from South America and one from South Africa.

BAIRD'S WHALE, plate nr. 83.

species	*Berardius bairdi* Stejneger, 1883.
area	North Pacific Ocean, cold and temperate (Chart 9).
sea temperature	5 to 20° C.
length	35 to 40 feet (10.7–12 m). Female about one foot longer than males.
weight	± 12 tons (700 lbs/ft).
food	Not recorded, probably Fish and Squid.
teeth	Two pairs in the lower jaw, a large pair (3″ = 75 mm) in the tip which extends about 4″ (10 cm) beyond the tip of the upper jaw, and the second pair about 8″ (20 cm) further back 2″ (50 mm) or less high. The teeth wear off with age.
illustration	From photographs. The colour is black or dark slaty grey, but looks lighter because of the numerous white scars, caused by fighting amongst each other. There are light patches under the throat, between the flippers (sometimes connected to one patch) and around the navel. The throat grooves have a length of about 2,5 feet (70 cm).
special features	The dorsal fin is 12″ (30 cm) high, the flukes have a spread of 12 feet (3.7) which is 31% of the body's length. The long cylindrical body, round head and conspicuous snout are characteristic. The flippers are very small. The head 13,5–23 % of the total length.
speed	Not recorded, probably cruising at 2 or 3 knots.
breath	Probably for several minutes regular breathing on the surface with 10 to 20 seconds interval, then sounding for a long period.
schools	No records, probably small.
biotope	Coastal and marine from Japan to Alaska, occasionally in the eastern Pacific Ocean down to Canada and USA.
migration	During the summer to warmer waters near Japan, in the winter north to the Aleutian Islands and the Sea of Okhotz. This seems the wrong way round but the Arnoux' Whale

does the same thing in the south and some other species perhaps also.

immatures   Pairing takes place in February and spring, the calves are born in December and have a length of 14 to 15 feet (4.3–4.6 m).

notes   This whale is caught in Japan by local whaling stations to ± 300 a year, all of them during the summer. 70% are males which seems to indicate that males and females live in separate schools and in different areas. In the Japan Sea mainly young animals are caught.

The photographs and much of the information which enabled the author to make the illustration and text, he received from the Whales Research Institute, Tokyo in the form of a publication by Professor Hideo Omura, K. Fujino and F. Kimura.

genus   HYPEROODON Lacépède, 1804.
NORTHERN BOTTLENOSED WHALE, plate nr. 84.

species   *Hyperoodon ampullatus* (Forster, 1770).
[= *Hyperoodon rostratus* (Müller, 1776)].

area   North Atlantic Ocean, arctic to temperate (Chart 5).
sea temperature   0 to 15° C.
length   Males to 33 feet (10.0 m), females to 27 feet (8.2 m).
weight   6 to 9 tons (450 to 600 lbs/ft).
food   Fish, Herring and Squid.
teeth   One pair in the tip of the lower jaw which does not extend beyond the upper jaw, height $1\frac{1}{2}''$, length $\frac{3}{4}''$, width $\frac{1}{2}''$ (3.2 × 1.9 × 1.2 cm). In females the teeth are smaller. Sometimes there is a second pair of teeth behind the first pair.

illustration   From life, photographs and cast. The colour is black to dark sepia-brown. With age this colour becomes lighter, first with yellowish white spots and areas mainly on the flanks and underside via a marbled stage eventually to almost totally yellowish white.

The melon of the head is round and bulbous and grows increasingly with age, looking like a balloon in old males. In the illustration the colour should be lighter for the size of the forehead portrayed.

special features	The body is heavy-set and round, circumference equals the length. The big round head is conspicuous. This whale rises fairly high out of the water when surfacing and sometimes cruises along with the head and back partially exposed.
speed	Cruising 2 to 3 knots.
breath	Many times during several minutes on the surface, then sounding for 10 to 20 minutes. They can remain under water for 2 hours. The blast is 2 feet high and noisy.
schools	Small, often solitary, 2 to 3 animals and occasionally 12. Round the Far Ør Islands there used to be thousands. After mating females and calves form separate schools from the males.
biotope	Marine and oceanic, deep water.
migration	North in the spring, returning south at the end of July, going as far as the Azores and occasionally into the Mediterranean. The majority stays north of 50° N and they are rarely seen on the normal shipping routes.
immatures	Pairing takes place during the spring when they go north and the calves are born a year later between March and May. The length is ± 10 feet (3 m).
notes	When harpooned this whale goes into a vertical dive to great depths, 600 fathoms have been recorded. At the beginning of this century as many as 3000 were caught yearly. This has decreased to only about 200 in 1964. On the 20th Aug. 1958 a live Bottlenosed Whale of 24 feet became trapped inside the sunken wreck of a coaster near Flushing in the Netherlands. There it was killed with a knife by the diver who was sent down to investigate the damage before an effort was made to lift the vessel. This position was far away from its normal habitat, although they are known to use the North Sea occasionally on their way south. The head contains a considerable amount of spermaceti (see Spermwhales).

SOUTHERN BOTTLENOSED WHALE.

species	*Hyperoodon planifrons* Flower, 1882.
area	Antarctic to temperate southern oceans (Chart 5).
sea temperature	0 to 15° C, occasionally higher.
length	Males 25 feet (7.6 m), females 22 feet (6.7 m).

weight	3 to $3\frac{1}{2}$ tons (320 lbs/ft).
food	Squid.
teeth	One pair in the tip of the lower jaw, $2\frac{1}{2}''$ (65 mm) high, showing $1''$ (25 mm) above the gum. The shape is conical with a diameter of $1\frac{1}{2}''$ (3.2 mm). Females have one pair of teeth in the same position, but they do not erupt.
illustration	As Northern Bottlenosed Whale, but the dorsal fin is placed further back.
special features	See Northern Bottlenosed Whale.
speed	Cruising probably 2 to 3 knots.
breath	Many times during several minutes on the surface, followed by a deep dive of 10 to 20 minutes. Can remain under water for 2 hours. The blast is short and broad, 2 feet (60 cm) high.
schools	Usually small, 2 to 12 animals but the author saw records of what must have been these whales, in the Humboldt Current off Chili of an estimated school of 40 (May, 1965).
biotope	Marine, the few records seem to give indication of a preferred distance of the coast of 100 to 200 miles (Chili and Bouvet).
migration	During the winter to warmer waters. The Antarctic records are from January, those on the coast of Chili from May.
immatures	No records. Probably mating in early spring, with the calves born a year later in warmer waters.
notes	A rare whale. Since 1882 only 10 strandings have been reported: 2 in Australia (Dec.), 3 in South America, 2 in New Zealand and 2 in the South Orkneys (Jan.), 1 in South Africa (Dec.). Records from whaling ships, however, indicate that they meet this whale not infrequently on the whaling grounds. The mentioned observation in Chili was at $25\frac{1}{2}°$ s. with a water temperature of $16°$ C. The observer reported "Pilot Whales" but from his description not black, colour dark brown, some greyish and several having large white areas on the body, it is clear that he observed the S. Bottlenosed Whale.

This family is divided into two sub-families:
Physeterinae : Genus *Physeter* – one species and
Kogiinae   : Genus *Kogia*   – two species.

The first genus was created by Linnaeus in 1758 as one of the few whales which were known to science. It is the only toothed whale of "whale size" and was then already commercially hunted for about 40 years.

The book "Moby Dick" by Herman Melville, written in 1850 is a novel about a dangerous and elusive white Sperm-whale. This book describes this trade and the species in more detail than many more modern works.

Sperm Whales could be approached by the sailing ships of those days and attacked with the small whaleboats. When killed they floated and could be flensed alongside the whaler, making them an acceptable prey long before the bigger and faster Finners, which sank after death.

Apart from the normal amount of whale-oil (average 6 tons per whale) obtained from boiling the blubber (33 % of the body weight in Sperm Whales) they yielded up to one ton of the precious spermaceti. This is a colourless oily fluid stored in a reservoir in the bulbous head. It can be separated in a high quality clear oil and a white wax. Both are used as ingredients in pharmaceutic and cosmetic industries. The wax was formerly used for production of candles.

The early discoverers of this oil, which is present in the heads of all toothed whales, especially in those with a bulbous melon, mistook it by its appearance for an accumulation of the male animal's sperm cells and named the species after it. Although this conclusion proved to be incorrect the name remained. The true function of this oil is not yet clear, but it is almost certain it fulfills a task in the body when diving to great depths.

The second genus is a much smaller, but similarly shaped animal, mainly known from stranded specimens all over the world. Only in Japan they are occasionally caught at a rate of 10 to 20 a year by the local whaling stations. The family feature is a rod-like lower jaw with many teeth, which is considerably shorter and thinner than the upper jaw and very small compared to the size of the head.

genus	PHYSETER Linnaeus, 1758.
	SPERM WHALE – CACHELOT, plate nr. 85.
species	*Physeter catodon* Linnaeus, 1758.
	[= *P. macrocephalus* Linnaeus, 1758].
area	All Oceans, females not beyond the sub-tropics (Chart 10).
sea temperature	Males 0 to 30° C, females 20 to 30° C.
length	Males to 65 feet (20 m), females to 40 feet (12 m).
weight	Males 60 tons (0,9 t/ft), females 18 tons (0.45 t/ft).
food	Squid and Octopus, Fish including big Sharks. It can swallow a man. Stomach contents quoted 75 to 100 squid ± 1 ton/day.
teeth	Variable 24 to 30 in each lower jaw, strong and conical. The crowns fit in sockets of the upper jaw. Length of the lower jaw is up to 15 feet, small rudimentary teeth in upper jaws.
illustration	From life and photographs. The colour is usually quoted as black, for the live animal at close quarters the author judges it dark sepia brown. Some specimens have been reported lighter underneath. Old bulls can have grey areas on the head and creamy-white albinos occur. The inside of the mouth is white, the lips often marbled. Scars round the head bear witness to their fights with giant octopus.
special features	The head is enormous and takes up $\frac{1}{3}$ of the body measuring 18 × 9 × 6.8 feet (6 × 3 × 2 m). The blunt almost square forehead is very conspicuous and always shows first on surfacing with the S-shaped blowhole 1'10" (55 cm) long, on the left side in the tip, exhaling a blast of 45 feet (15 m) high. This blast has an inclination of about 45° forward, which is diagnostic for the species because all other whales have a vertical blast. The dorsal fin is a laterally compressed hump. Many illustrations give a series of humps along the tailstock, the author noticed only 3 just in front of the flukes. They often lie quiet on the surface blowing, with the body exposed from the blowhole to the hump. When sounding the tailstock and the flukes which have a spread of 16 feet (5 m) always show in the air.
	The dive is a vertical one. During the mating season (April and May) both sexes sometimes jump right out of the water, almost vertical, falling back in any direction with a terrific

151

splash. Occasionally this is seen in the summer as well and they practise lobtailing with the whole tailstock out of the water.

speed	Cruising 3 to 4 knots but often quite still on the surface. Capable of speeds up to 12 knots.
breath	On the surface for about ten minutes, females blowing every 10 seconds, males every 20 to 30 seconds, then sounding for a dive which can last from 5 to 80 minutes.
schools	Normal 3 to 5, often 1 to 2 bulls with several cows.

The schools are spread out over a distance of 10 to 40 miles or more. The large schools of the last century are no longer seen. The largest one the author saw was in the Indian Ocean and had 10 animals.

biotope	Coastal and marine in deep water, on migration also oceanic. Rarely seen further away than 300 miles from shore or island. Typical localities are Sth. coast Ceylon, Arabian coasts of the Indian Ocean, Maladives, Philippines and Galapagos Islands.
migration	There is a definite migration to higher latitudes in the summer, extensive for males, only partly for females.
immatures	Pairing takes place in the spring, April and May, when the spectacular leaps can be seen on the normal shipping routes through the Philippine Islands and the Indian Ocean.

The calves are born after 16 months (according to Slijper), other zoologists quoted 10 months. They are a little less than half the length of the mother who nurses them for $\pm$ 12 months (or 6 months) until they have reached a length of 20 to 22 feet (6–6.7 m). Females are mature after 15 months and males after 2 years. Other sources quote $4\frac{1}{2}$ years for both sexes.

notes	These whales are known to dive to at least 3240 feet, because one adult was found drowned at that depth off the South American Pacific coast, tangled up in 180 feet of deep sea telegraph cable.

In the heyday of spermwhaling as many as 10.000 a year were caught, this decreased sharply around 1850. In modern times these whales are hunted during the off-time on the whaling grounds before the "season" opens, because there is no restriction for this species.

Before World War II this amounted to about 1000 a year. In more recent times, with a fast dwindling stock of Finners more attention was diverted to Sperm Whales. In the 1953 season 8317 were killed. See table page 167. The blubber, which is 33% of the body's weight, reaches a maximum thickness of one foot (30 cm) between the flippers. For spermaceti see introduction to this chapter.

From the stomachs and intestines of sick Sperm Whales comes another very valuable product, the famous Ambergris, a whitish grey waxlike substance which has a musky odour. It floats on the water. Because of its superb odour-retaining quality it is an essential ingredient of all perfumes, essences and expensive soaps. The value is 200 pounds sterling per lb. The biggest piece ever found was 1000 lbs.

genus	KOGIA Gray, 1846. PYGMY SPERM WHALE and DWARF SPERM WHALE, plate nr. 86.
species	a. *Kogia breviceps* (de Blainville, 1833). b. *K. simus* Owen, 1866 is described as a separate species, but not enough external data are available to give a separate description.
area	Tropical and subtropical oceans, occasionally temperate (Chart 6).
sea temperature	Above 20° C.
length	a. 10 to 13 feet (3.0–4.0 m); b. 2.1–2.7 m.
weight	± 1500 lbs (680 kg).
food	Squid and Crabs.
teeth	a. 2 × 11 to 15 in the lower jaw only; a recently stranded specimens in New Zealand had 2 × 16. b. 2 × 9 to 11. The teeth are long, slender, sharp and incline backward.
illustration	From photographs, different casts and life.
special features	The same typical "square" head as the Sperm Whale, but smaller and more variable in shape. The blowhole is above the eye and a little left of the middle. The dorsal fin is developed as such and not just a hump. Flukes and flippers are dark on both sides.

speed	Cruising probably below 4 knots but capable of much more (notes).
breath	No records.
schools	No schools have been reported. This rare species certainly does not congregate in large schools, probably single or with 2 or 3 together (notes).
biotope	Deep water, probably marine.
migration	No migratory movement can be deduced from the recorded strandings. They are more numerous in the western sides of the N. Atlantic, N. Pacific and N. Indian Oceans than on the eastern sides and south of the equator.
immatures	In two stranded females, foetuses of 9″ and 8″ were found, respectively in December and January (23 and 20 cm); a foetus in March was 3′7″ (109 cm). The calves are probably born in late spring and have a length of $4\frac{1}{2}$ feet (137 cm). Nursing probably lasts 12 months.
notes	The author observed what he considers to have been these whales on several occasions in the western Indian Ocean between Sokotra and the Arabian coast. Singles or in pairs, jumping the same way as Sperm Whales vertically out of the water falling back from the upright position. They were too shy to be approached closely.

In 1964 an adult specimen stranded near Napier, Hawkes Bay, New Zealand, still alive and was successfully transported to one of the basins of Marineland there. When released it charged the men of the basin with snapping jaws. Both men were hanging on the sling and were pushed by the whales' head right out of the water. It then circled round the tank with such force that the bow wave came over the edge of the pool. Later it became more docile and loved to be scrubbed with a hard brush, but refused food and died within a week purposely ramming its head against the wall of the tank. Section revealed that it was very ill internally. This supports the previously stated opinion that mainly old and sick animals near the end of their days, beach themselves perhaps to avoid drowning or falling victim to sharks. To science they are only known from stranded specimens, these are indicated in the chart, the majority in temperate waters.

All (20) strandings in New Zealand occurred in sheltered bays of Cook Strait, Hawkes Bay and Hauraki Gulf.
In Japan they are included in the whale catches of the smaller local whaling factories (4 specimens in 1952).

WHALEBONE WHALES
Mysticeti

FAMILY: BALEEN OR RIGHT WHALES (*Balaenidae*).
There are two genera: *Balaena* with 2 species
and: *Caperea* with 1 species.

Some zoologists consider the species *Balaena glacialis* as a separate genus: *Eubalaena*.

The distinguishing features of this family are the absence of grooves in the throat and a back without dorsal fin.

Genus *Caperea* conforms with genus *Balaena* but has a normally developed dorsal fin. They are extremely rare and have only been recorded from the southern oceans.

Genus *Balaena* are stout and heavy whales, slow moving, stupid and defenseless. Because of this the pursuing whalers gave them the title of being a "right" whale and they were the first to be slaughtered for the comforts of mankind.

Whale-oil was for many years the only means of lighting, and whalebone, before the time of steel and plastic was worth a fortune. At its peak it brought 2250 pounds sterling per ton., which decreased to 700 pounds sterling and 400 pounds sterling. A full grown whale yielded $1\frac{1}{2}$ tons.

Whalebone is a horny substance growing downwards as plates right and left in the mouth from the crosswise ridges in the palate. They are triangular in shape, ending in a point near the lower jaw on either side of the enormous tongue.

The inside edges of the plates are fringed with horny hair or bristles, which, when the tongue is pressed upwards on closing the mouth, forms a sieve straining the plankton from the seawater, which is squirted back into the sea between the lips.

These bulky animals float higher on the water than other species. A Right Whale weighs about 220 % more than a Sei- and 135 % more than a Spermwhale of the same length.

Of the body-weight approximately 45–36 % is blubber, 30–35 % meat, 11–14 % internal organs and 14 % bones, compared to 33–43–9.5–10 % in

Sperm and 18–58–9–13 % in Sei.

Porpoises have 45–60 % blubber, dolphins 30–45 %. The smaller the animal, the larger is the surface of the skin compared to the weight and more heat is lost by radiation, which is regulated by the blubber.

An elephant has more than 4 times the skin surface of a Blue Whale in relation to its weight. Water takes up heat 23 times faster than air. The bigger and rounder the whale therefore the more economical is the "heating system".

The Netherlands Whaler Willem Barentsz had to shoot a Right Whale in 1958 in the Antarctic (with special permission) to serve as floating bumper between herself and a refuelling tanker. Only dead whales can serve for that purpose. When the operation was finished the whale had to be partly flensed in the water because it was too big to pass through the gate in the stern of the ship.

Once in the North Atlantic Ocean, east of New Foundland Banks, the author spotted three of these whales on the three cm radar at a distance of 3 miles. The visibility was doubtful and the radar was operating in order to check it by means of passing ships.

The three echo's appeared but no ships could be detected in that direction. Measures were taken to reduce speed until the intensive search with binoculars showed the three blowing Right Whales, proving the visibility was much better than estimated.

Right Whales are protected by the International Convention for the Regulation of Whaling. Earlier conventions imposed restrictions in 1929 and completely stopped whaling of this species in 1936.

In the heyday of southern whaling, when the ships in search for Sperm Whales found these species in abundance, 14.000 Black Right Whales were slaughtered per year in the New Zealand zone. During the season 1934–'35 only 4 were caught, 2 in the North Pacific and 2 near South Africa.

There are now signs of recovery. They have been reported as "quite numerous" (6 seen daily by fishermen) round Tristan da Cunha (South Atlantic) and St. Paul (Ind. Ocean) and Campbell Island (N.Z.) Recorded sightings by Japanese whalers in the North Pacific were 6 in 1953, 37 in 1954, and 70 in 1957. In the North Atlantic Ocean 37 were observed from 1954 to 1956.

The population round Campbell Island (N.Z.) was estimated at 100 in 1968 and one sub-adult recently visited Wellington Harbour. The purely Arctic Greenland Whale was last harpooned in 1899 and considered to be extinct. Reports however filtered in that the Eskimos of Northern Alaska still occasionally saw and caught it. More recently it has been observed in Baffin Bay again.

Its original "home waters" were from Spitsbergen round Greenland to the

Canadian arctic. Whaling drove it back into the icefields of Baffin Bay already by 1720 roughly 100 years after the pursuit started. Another 100 years finished it there as well. The last efforts were concentrated on the Bering Sea area.

During recent counting of whales in and north of Bering Straits 5 or 6 records of probably Greenland Whales appear.

It is not generally realized that seawater temperatures in that polar area rise to 5 and 10° C during the summer months.

genus	BALAENA Linnaeus, 1758. [= *Eubalaena* Gray, 1864]. BLACK RIGHT WHALE, and SOUTHERN RIGHT WHALE, plate nr. 87.
species	*Balaena glacialis* Müller, 1776. *Balaena g. glacialis* in the North Atlantic Ocean. [= *B.g. japonica* Lacépède, 1818]. [= *B.g. sieboldi* Gray, 1864]. [= *B.g. australis* Desmoulins, 1822]. and several others considered to be different geographical names for the same species.
area	Temperate – plankton rich waters Atlantic – Indian- and Pacific Oceans (Chart 6).
sea temperature	North 10–20° C, South 5°–20° C occasionally 25° C.
length	58 feet (17.6 m), spread flukes 41% of total length. Head $12\frac{1}{2}$% of total length.
weight	With length 40 feet 24 tons (0.6 t/ft) length 58 feet 72 tons (1.25 t/ft).
food	Plankton.
whalebone	To 250 pairs of black plates, max. length 9 feet (2.7 m).
illustration	From photographs and life. The colour is black, which in the North Pacific Ocean is reported to be blue-black and in the southern oceans purplish-black. The belly sometimes has small white areas, this seems to occur more (one in five) in the south than in the north. These white areas have pink edges where they join the black.
special features	A big fat whale, showing more above the surface than other whales, very slow and without a dorsal fin. They exhale a double blast from two nostrils and always dive vertically showing the flukes. On the tip of the snout is a conspicuous horny growth, called the bonnet, which is usually discoloured by parasites.
speed	3 to 4 knots, maximum 7 knots.
breath	5 or 6 times lying on the surface, then a deep dive which can last to 10 or 20 minutes. The blast is 15 feet (4.6 m) high.
schools	Single or in pairs (97%) never more than 4 together. (A school of 22 reported in the Cape Verde Islands seems therefore rather unlikely and in the author's opinion might

159

have been Humpback Whales, which look, blow and move in much the same fashion and often do not have a very distinct dorsal fin).

biotope   Coastal and marine, occasionally pelagic on migration. They seem to prefer the neighbourhood of islands or island groups.

migration   To colder climates during the summer returning in autumn.

immatures   Calves are born in the winter and stay with the mother one year.

notes   Hunted almost to extinction but recovering. They are infested with parasites especially on the bonnet, round the blowholes and in the mouth.

The author nearly rammed one south of Gough Island in 1936. It was obviously fast asleep on the surface and woke up when the bowwave broke over its head.

Right Whales are reported to be favourable preys for the much feared Killer Whales, which attack them in packs and seem to have a special preference for the huge $1\frac{1}{2}$ ton tongue. The thickness of the blubber is 6 to 12" (15 to 30 cm).

The photographs and information which enabled the author to make the illustration are from Prof. Dr. E. J. Slijper's "Whales" and from a publication on this species by Prof. Hideo Omura of the Whale Research Institute in Tokyo, which was presented to the author for this purpose.

GREENLAND WHALE, plate nr. 88.

species   *Balaena mysticetus* Linnaeus, 1758.

area   Arctic waters from Alaska to Greenland (Chart 6).

sea temperature   0 to 5° C.

length   60 feet (18.5 m) maximum.

weight   Probably 110 tons (1.8 t/ft). Weights quoted from 200 to 300 tons are almost certainly exaggerated.

food   Plankton.

whalebone   300 to 360 pairs of black plates, length 13 to 15 feet (4.6 m).

illustration   From description and other illustrations.

The only good description came from William Scoresby in 1820. This species was slaughtered practically to extinction during the 18th and 19th century and no authentic scientific data about it are available. The head takes up $\frac{1}{3}$ of the body with a mouth 20 feet deep and 10 feet wide (6.1 $\times$ 3.0 m). The flippers measure 8 by 4 feet (2.4 $\times$ 1.2 m). Flukes have a length of 6 feet (1.8 m) and a spread of 26 feet (8 m) which is 43% of the total length. The blowholes are 6 to 8″ (15-20 cm) and the thickness of the blubber is 1′8″ (50 cm).

special features	A thickset heavy whale without dorsal fin and showing a double blast. The upperlip, eyelids, tailstock and flippers are reported light grey, the lower jaw yellowish white, the rest of the body black. They dive vertically showing the flukes and return to the surface in the same spot.
speed	Probably 2 to 3 knots.
breath	On the surface 4 to 6 times during 3 minutes, then a deep dive for 10 to 15 minutes but they can stay under water up to 1 hour. The blast is 12 to 20 feet high (3.7–6.1 m).
schools	When they were numerous they were found in schools.
biotope	Always amongst or close to arctic pack-ice. Formerly known from Bering Sea throughout the Canadian Arctic via west and east coasts of Greenland to Spitsbergen. Probably coastal.
migration	Pairing takes place in early spring. The calves are born after 9 or 10 months, have a length of 10 to 15 feet (3–4.6 m) and are nursed by the mother for one year.
notes	This species has no external parasites. Needless to say it lives far outside the normal shipping routes.

genus	CAPEREA Gray, 1864. [= *Neobalaena* Gray, 1874]. PYGMY RIGHT WHALE, plate nr. 89.
species	*Caperea marginata* (Gray, 1846).
area	Temperate southern oceans, perhaps subantarctic (Chart 6).
sea temperature	5 to 15° C.
length	± 20 feet (6.1 m).
weight	± 5 tons.
food	Plankton.
whalebone	230 pairs of plates, maximum length 2′3″ (70 cm). They are ivory white with a dark outer edge and have a very soft fine fringe.

illustration	Description and photograph of a preserved calf. The head is $\frac{1}{4}$th of the total length. The flukes of a 9 feet specimen (2.8 m) had a spread of 20% of that length. The colour is black with a narrow white line along the belly. Immatures are dirty white underneath.
special features	A small sickleshaped dorsal fin behind the middle.
speed	No records.
breath	No records.
schools	Probably single or in pairs.
biotope	Probably marine and oceanic.
immatures	Newborn calves have a length of 9 feet (2.75 m) and are probably born in spring.
notes	An extremely rare whale. The nominate specimen probably stranded in New Zealand in 1846. 12 possible strandings have been reported there since then. The last one in 1951. During spring storms 3 stranded on Kangeroo Island in the Australian Bight; the newborn calf date unknown. – one of 16 feet (4.96 m) 28th Oct. 1884. – one of 11 feet (3.35 m) 21st Oct. 1889. and one in Bass Strait June 1929. From South Africa 2 are known, both stranded in False Bay, east of Cape Peninsula January 1917 one of 11 feet and another one in January 1937, measurements unknown because it was already burned on the municipal rubbish dump before the scientist heard about it. This species has 2 pair of ribs more than the maximum of 15 in other whales.

GRAY WHALES (*Eschrichtidae*)

This family has one genus: *Eschrichtius* with one species living in the North Pacific Ocean. A separate population inhabited the North Atlantic Ocean but became extinct about 1725.

The Gray Whales are intermediate between the Right- and Finner Whales, primitive and if it can be expressed that way, nearer to their previous land-ancestry than the other species.

They are extremely coastal, frequent small bays and shallow waters, play in the surf and sometimes remain stranded during the low tide in no more than 2 to 3 feet of water without panic or injury.

They also have more hair than the other whales, growing on top of the snout and along the lower jaws. In the throat are two expansion grooves, occasionally four against none in Right Whales and many in the Finners. They have no dorsal fin but a series of low humps.

The original stock in the North Pacific Ocean was estimated at several hundred thousands. Through indiscriminate slaughter by coastal whaling during the 19th and beginning of this century it was reduced to so few that the Eastern Pacific population was considered to be extinct by 1911. They reappeared however by 1925 and became a protected species when International Whaling Interests agreed upon measures to save rare and nearly extinct groups. They recovered considerably since then and the present eastern stock is estimated to be about 6000.

The only region where they are professionally hunted is Japan where local whaling stations chase them in little boats as they have done for over 200 years.

As is indicated on the chart there seem to be two populations, one on each side of the Pacific. It is thought by Japanese whale authorities that they do not mix. The western stock was regarded as practically extinct by 1966.

genus	ESCHRICHTIUS Gray, 1864. [= *Rhachianectes* Cope, 1869]. GRAY WHALE, plate nr. 90.
species	*Eschrichtius gibbosus* (Erxleben, 1777). [= *E. glaucus* Cope, 1868].
area	North Pacific Ocean (Chart 4).
sea temperature	5 to 20° C.
length	To 45 feet (13.7 m). Females larger than males.
weight	± 20 tons (1000 lbs/ft).
food	Plankton. They are reported to eat only when in summer quarters.
whalebone	150 pairs of plates, thick and strong, maximum length 16″ (46 cm), yellowish white colour.
grooves	Usually 2, sometimes 4, length ± 5 feet (1.5 m).
illustration	From photographs. An ugly ungainly animal with thick flippers and flukes which are often damaged. The grey body is discoloured by many parasites. On the back and tailstock are 6 to 14 irregular humps and no dorsal fin. In older specimens these humps are often damaged too.
special features	No dorsal fin. The irregular skin infested with whale lice and barnacles. They sometimes stand "on the tail" vertically in the water, head out to look round. They do this amongst the ice floes in the north as well as in winter quarters. They move and dive slowly and when sounding show the flukes.
speed	2 to 4 knots, maximum 6 knots.
breath	On the surface 2 or 3 times in succession, then a deep dive lasting 8 to 10 minutes. The blast is about 10 feet (3 m) high but because they exhale rather slowly, not very conspicuous.
schools	Last century of 1000 and more, recently back to 40 and 100. Except on migration; the females with calves form separate schools.
biotope	Coastal, very close inshore, preferring shallow bays and inlets. Plays in the surf.
migration	During summer in Sea of Okhotsk and Bering Sea as far as the floating ice, in autumn and winter southwards to Japan and Lower California, travelling close inshore.
immatures	Pairing takes place in the winter, the calves are born after one year, one or two. Length 16 to 18 feet (4.9–5.1 m), weight 3400 lbs (1.5 tons). They are nursed by the mother

for 6 to 8 months until they have reached a length of 25 feet (7.6 m) and stay with her for one year.

Scientists tried in 1953, 1954 and 1956 to take a cardiagram of these whales. During the last effort a mother and calf were approached with an 18 feet speed-boat. The calf showed fright and in consequence the mother attacked the boat and battered in one side approaching from underneath. These whales are prey to Killer Whales and when meeting them are said to be paralysed with fright.

The blubber has a thickness of 10″ (25 cm).

This family has two genera: *Balaenoptera* with 5 species
                           Megapterà   with 1 species

The members of this family are long and slender with pointed snouts. They are fast swimmers and have a varied number of grooves in the skin of the throat and chest. These allow expansion of the mouth when feeding to hold a maximum of seawater with krill and plankton, as well as for the chest when breathing. The shallower diving species have less and shorter grooves than those who feed in deeper water. Embryo Finners still have small teeth in the lower jaws.

These whales show little of the body when blowing. They rise slowly at an angle of about 20°, which brings the head from snout to blowhole horizontally touching the surface to blow, then a little more of the head appears before it dips down again followed by the slightly bent back with the dorsal fin placed far back, finishing with the arched tailstock.

The faster they swim the more they show, both of the head when coming up and of the body when diving because it is arched more strongly.

Not until the harpoon-gun had been perfected and faster, larger chasers were used, could these whales be hunted with success. The harpoon needed a device which blew air into the killed whale because they sink when dead. Their enormous power is demonstrated by the fact that one harpooned Fin Whale took the chaser working her engines half astern in tow with a speed of 8 knots for several hours before it could be finished off. In spite of this power and their size, they are completely defenseless and there are no records of any of these species ever turning on and attacking their pursuers.

The table reveals some facts about the effects of modern whaling up to the present day. The figures are taken from books and publications. In the early thirties 90 % of the whales were caught in the Antarctic, 80 % were Blue- and 20 % Fin Whales.

Sei-, Humpback- and Sperm Whales were commercially unattractive as long as the bigger ones were in good supply and they were mainly taken by the land stations.

The yearly catches of Gray- and Right Whales in the 19th century reduced the bag of Gray's to about 5 in 1926 and that of the Rights to 10 in 1935 (6 in the N. Pacific and 4 in the Antarctic).

These two species were practically exterminated.

year	total	blue (1)	fin (2)	sei (6)	humpb. (2,5)	sperm	other
1930/31	37461	± 26000 80 % ±	6600 20 %	—	(3500)	(900)	6 RW
1933/34	32586	—	—	—	—	—	—
1934/35	44855	22500	16500	466	3465	924	10 RW
1937/38	54835	17700	33600	196	2279	967	—
1946/47	43669	8870	12857	2	antarctic only		
1952/53	—	4208	25553	2172	3322	8317	97 M
1956/57	21729	1500	25000	—	1250		
1963/64	28921 - antar.	112	13870	8286	2	6651	101 M
	18300 - N. Pac.						
	22549 - elsewhere				(1000-4000)		
					illegal		
	69770 - total						
1964/65	31413	20	7.308	19.874	0	4.211	—
1965/66	24441	1	2.318	17.583	1	4.538	
1966/67	20225	4	2.893	12.368	—	4.960	
1967/68	15080	—	2.155	10.367	—	2.568	

In the last season before World War II the percentages for Blue- and Fin Whales had changed from 80 %–20 % to 34 and 66 % and in the first season after the war the Blues had recovered to 41 % against 59 %. The period of rest had been a help, but was not enough. By 1953 it became 14 %–86 % and with less than 1 % in 1965 it is obvious that the Blue Whale has become a very rare animal. Concurrent with the decrease in the catch of bigger whales, that of the smaller species and Sperm Whales had dropped in results from 100 % in 1952 and the previous years to 70 % in 1961 and 41 % in 1964. ("Hundred years after Svend Foyn" by Prof. Dr. E. J. Slijper).

Until the thirties whaling was a free enterprise. Norway as one of the biggest whaling countries realized already at this stage that the industry was in danger of killing "the hen which laid the golden eggs" and imposed in 1929 a national restriction protecting immatures, mothers with calves and the Right Whales. In 1936 this was followed by an international conference on whaling, resulting in the Geneva Convention which adopted similar restrictions on an international scale, installing "Inspectors" on the whaling fleets to see to its enforcement. Several of the largest Whaling-countries, however, did not join the convention. After World War II it was clear that something better had to be done to save the whales and the industry. A conference was held in Washington in 1946 where the International Whaling Commission (I.W.C.) was founded with a secretariat in London and an International Bureau of Whaling Statistics in Norway.

Antarctic whaling which could only be done during the summer months was further restricted to a season commencing January 7th and ending on the date when the quota of allowed "Blue Whale Units" (BWU) had been caught. This was usually around the middle of March.

A BWU is one Blue Whale, for the other species the number ( ) in the table indicates how many of that species make up 1 BWU.

The main source of the whale oil is the blubber. In the following list the % are given how much blubber the different species have of their total weight. Blue: 22–27; Fin: 21; Sei: 18–20; Humpback: 33; Sperm: 33; Right: 36–46; Dolphins: 30–45; Porpoises: 45.

The amount of oil obtained from the blubber is about half that weight.

To make the decision on the closing date of the season possible all whaling factories have to send in their catch records weekly to the IBWS. Before and after the season only Sperms are allowed to be caught, also with a restriction to size which practically protects the females.

In addition to season and size there are geographical restrictions which it is hoped will work as game reserves.

From the gathered statistics the IWC each year decides how many BWU can be caught.

Even in 1953 this decision was disputed and some whaling interests produced their own statistics which indicated that a double quota could be killed without harming the stock.

To try and solve this problem the IWC requested the merchant navies of several countries to make and record observations of Whales at sea. This was done with great enthusiasm from 1955 to 1958. Before the results of this count could be worked out the actual drop in catches already proved the IWC right and the quota even too generous. This quota varied from 16.000 in 1953 to 15.000 in 1963 and 10.000 in 1964 and it is significant for the situation of the stock that in spite of all efforts, modern equipment and efficiency in 1963 only 11.306 BWU and in 1964 8.429 BWU were caught.

In the following paragraphs for each species the figures quoted are all from Slijper's publication.

BLUE WHALE: It is obvious that this whale is nearing the end of its existence unless drastic restrictions are imposed and adhered to.

Statistical calculations based on the antarctic whaling results indicate that of an original stock of 100.000, in 1963 no more than 1.000 to 1.500 were left.

Only 100 % protection can save them but this can not be imposed. They are now only allowed to be hunted between 40 and 50° South and 00° to 80° East where the Japanese discovered a new species called the Pygmy Blue Whale. The figures for 1965–1968 show that this was probably only wishful thinking.

FIN WHALE: For this species the calculations show that the original stock was about 200.000, reduced by 1963 to 40.000. From this remaining stock 4.000 could be killed yearly without endangering the continuation of their existence but this restriction could not be imposed either. The quota for 1964 was still 20.000 specimens (with no Blue Whales available) and only 14.000 were caught. Only half that number could be obtained during the next season followed by a further spectacular reduction in 1966. This clearly indicates that the Fin Whale too is on the brink of extermination.

SEI WHALE and BRYDE'S WHALE: these smaller whales have not been commercially hunted long enough to make an estimate on past and present stock. Before World War I the numbers killed yearly were probably between 200 and 500, nearly all by land stations.
After the war this increased sharply to over a thousand. Between 1949 and 1958 the average is 1646 for land stations.
The share taken by Antarctic whaling increased from 2 in 1947 to 2172 in 1953 and 8286 in 1964. Undoubtedly the land stations also considerably increased their catch during the last 15 years.
The following season with the Blue Whales practically extinct and Fin Whales reduced to $\frac{1}{3}$ of the catch allowed, the Sei Whale had to pay the bill and nearly 20.000 were taken.
These whales have the advantage of less commercial value and combine a great speed with a rather spread out subtropical and tropical distribution where they are comparatively free from pursuit. The steep increase in killings certainly asks for attention and consideration, especially since the number caught in the "all out" effort during 1968 was only half of that three years earlier.

HUMPBACK WHALES: From a yearly catch around 3000 mainly by land stations before World War II there has been a sharp increase in numbers taken which the figures in the table do not show.
Statistical calculations indicate that from an estimated stock of 15.000 in the Australian sector of the Antarctic they have been reduced to 1.000 or 1.500

whales by 1963. For this stock to recover, 50 years of "rest" would be necessary.

This measure could not be imposed but the hunt has been drastically restricted and is completely forbidden in antarctic waters. In 1964 the Australian and New Zealand land stations were closed down (see notes nr. 96).

It is to be regretted that observations show that still 1.000 to 4.000 a year were shot illegally.

These whales share a wide and partly tropical distribution with the previous species but their migrations were always along very coastal routes where with a relatively slow speed they served as "sitting ducks" for the catchers.

There might be signs that they are changing these routes because the author has observed them in recent years in very pelagic tropical areas where they were never seen before. On the voyage from Dar es Salaam to Luanda in July 1966 which passes through their traditional winter quarters on both sides of Africa not a single Humpback was seen.

For Nature Conservationists the most revealing, surprising and disquieting figures are from 1964, a poor year with only 41 % of the normal results in Antarctic whaling. Besides the 28.921 whales registered as killed there, which does not include the illegally shot Humpbacks and perhaps up to 10.000 Sperm Whales, the North Pacific whalers caught 18.300 whales and the land stations all over the world 22.549 whales. This makes an official total of 69.770, in reality perhaps as many as 75.000. Considering that this is about 50 %. more than the number which raised such anxiety that the Geneva Convention was signed, introducing voluntary restrictions, then the poor whales have not had much benefit from the protection.

New methods to extract oil from fish to cover the demand finally look like bringing about a change for the better.

The competition of "herring-oil" had become so strong in 1968 that whale-oil could not be sold at a paying price. In consequence the large Norwegian Whaling Industry decided to stop operating, leaving only Japanese and Russian whalers in the field.

The Japanese coastal stations have already so much exhausted the supply of smaller whales that they are turning to dolphins. Over 200.000 are now caught yearly. The problem of the Preservation of Nature in the world's oceans is therefore by no means solved but seems to be shifting.

Genus	BALAENOPTERA Lacépède, 1804. [= *Sibbaldus* Gray, 1846]. MINKE WHALE or LITTLE PIKED WHALE, plate nr. 91.
species	*Balaenoptera acutorostrata* Lacépède, 1804. [= *B.a. davidsoni* Scammon, 1872, N. Pacific]. [= *B. bonaerensis* Burmeister, 1867, Southern Oceans]. [= *B. huttoni* Gray, 1874, Southern Oceans].
area	All oceans more in temperate than in tropical waters (Chart 11).
sea temperatures	5 to 10° C preferred, also in tropical areas 25 to 30° C.
length	25 to 30 feet (7.6–9.2 m), females are 2 feet longer than males and they reach maturity respectively at lengths of 22 feet and 24 feet.
weight	± 10 tons (750 lbs/ft).
food	Krill and Fish (length to 12″: Sandlance, Sardines, Anchovy and Herring occasionally Squid.
whalebone	The Atlantic species 270–348 pairs of plates; in the North Pacific Ocean 230–295 pairs of plates. The colour is yellowish white with a fine white fringe. Max. length 10 to 11″ (27 cm).
grooves	50 to 70 extending to half-way the flipper.
illustration	Life and photographs. The white mark on the flipper is sometimes longer and can also extend over the full width. The white of the belly normally reaches to behind the flipper but sometimes up into the dorsal blackness, changing to grey on the back towards the spine where the grey of both sides meet. The N. Pac. race has a shorter snout and a lower number of plates. The *B. bonaerensis* is a different race or colour-phase. Dr. W. L. van Utrecht of the Zoological Laboratory in Amsterdam described another variation from the southern oceans in 1961. There is no sharp division between the black and white on the flanks; the flippers measure 17 % of the total length against 10 %–16 % in B.a. and the telltale white patch is absent.
special features	A small, stout, black and white coloured Finner with a well developed dorsal fin from which originated the name of Little Piked Whale. The white patch on the flipper is diag-

nostic.

They hover around fishing fleets waiting for spilled fish and never get mixed up in the nets. It is probably the only whale which sometimes accompanies big ships at sea swimming alongside 3 to 5 cables off and is reported to dive under the keel from one side to the other.

They jump out of the water quite often shooting out at a 45° angle as far as the flukes, falling back sideways or over backwards in the sea, usually 2 or 3 times in succession. The flukes have a spread of 28 % of the total length.

speed
To 16 knots when accompanying a ship but also fast when presumably pursuing fish. Cruising 2 to 4 knots.

breath
When on the surface 5 to 8 times with shallow dives followed by a deep one lasting from 5 to 10 minutes. Sometimes staying on the surface for a long time, continuing with shallow dives.

The blast is 6 feet high and noticeable in tropical temperatures.

schools
Schools of many hundreds were normal but they have been hunted so intensively in the North Atlantic Ocean over the last years that their numbers have decreased sharply.

In the tropics the schools vary from 2 to 10.

biotope
A coast-loving species but also marine and occasionally oceanic. In Japan all Minke Whales are caught within a range of 2 to 3 miles from the shore.

migration
During the summer to higher latitudes, returning to warmer waters in winter. Late departers sometimes get caught under the frozen polar seas and can survive only by keeping breathing holes open. When these freeze as well they must eventually suffocate. The males migrate first, followed by the females with weaned immatures in separate schools.

immatures
Pairing takes place from February till March in the north; from August till September in the south. Calves are born after 10 months and have a length of 9 feet (2.7 m), weigh ± 450 lbs (204 kg). They reach maturity after 2 years.

notes
About 400 are caught annually in Japan and ten times that number in Norway.

One specimen swam alongside the author's ship in the Bay of Bengal, sideways, the belly turned towards the ship and

172

part of the flukes above the surface. The first impression, mistaking the flukes for an exposed dorsal fin, was of an enormous white whale. Perhaps most of the animals seen in tropical waters are immatures. The author has seen them regularly round Ceylon, near Christmas Island (Ind. Oc.) on both coasts of tropical Africa (Mozambique and Angola), in the Bay of Biscaye and around the Galapagos Islands. The thickness of the blubber is $2\frac{1}{2}$ to 4″ (5–10 cm).

They are prey to the *Orcinus orca*. It is relatively rare in the S.W. Pacific. New Zealand recorded only 13 strandings in the period 1873-1968, these were all immatures.

	Sei Whale, plate nr. 92.
species	*Balaenoptera borealis* (Lesson, 1828). [= *B.b. borealis* Northern Hemisphere]. [= *B.b. schlegeli* Flower, 1864, Southern Hemisphere].
area	All oceans, rare in Australian and New Zealand regions (Chart 12).
sea temperature	5 to 30° C.
length	40 to 58 feet (12–18 m) females longer than males.
weight	To 23 tons ($\pm$ 0.4 t/ft).
food	Krill and sometimes Sardines. Stomach contents 450 lbs.
whalebone	340 pairs of plates, black with a very fine and soft white fringe. Maximum length 2′2″ (55–65 cm).
grooves	32 to 62, extending to 3 or 4 feet beyond the tip of the flipper which is half way between that tip and the navel.
illustration	From life and photographs. The chin, throat and belly as far as the anus are always white. From the flanks the darker dorsal colour invades this white in various shapes. In Bryde's Whales (nr. 93) this area is dark, in Fin Whales (nr. 94) the right side of the head is dark, the left side and the whole of the underside is white. In Blue Whales this is blue.
special features	A medium sized dark coloured Finner, the white on the underside will rarely be seen in the field. (A Fin Whale is light grey and a Blue Whale is very much larger and bluish). The undersurface of the flukes is also dark. When surfacing they first show the blast, followed by the head from the tip of the snout to the blowhole. When that submerges again

the long arched back appears with a "good-sized" dorsal fin, situated rather far forward for a Finner, then part of the tailstock before they disappear almost horizontally under water. They swim straight without rolling, do not jump out of the water except on rare occasions in the mating season when pairing and never show the flukes.

speed
: Cruising 5 to 6 knots, capable of spurts to 30 knots over short distances.

breath
: Blowing 2 of 3 times every 17 seconds, then a dive lasting 5 to 10 minutes. Also about 5 or 6 times with 35 to 50 sec. intervals and then a longer dive. The blast is 10 to 14 feet high (3 to 4 m).

schools
: Single or in pairs when travelling, 10 to 50 in areas with plentiful food, always spread out. Formerly there were thousands.

biotope
: Coastal and marine, usually in deep water but also close inshore (7–10 miles) in depths of 12 to 20 fathoms.

migration
: During the summer in cold temperate waters, migrating to tropical areas in the winter.

immatures
: Pairing takes place in late autumn and early winter. The calves are born after 12 months, one or two. They have a length of 15 to 16 feet (4.7 m), weigh 1.2 tons and are nursed for 5 months. When weaned they have a length of 26 to 30 feet (8–9 m) and reach maturity after 18 months for males at 44 feet (13.4 m) and for females at 47 feet (14.3 m).

notes
: These are the Finners which seafarers will often see because they frequent the usual coastal shipping routes.

They are not particularly shy of ships and when these are heading straight for them, they do not divert from their course or sound until the ship is very near. When the observer is lucky he can see them blow as close as 100 yards from the bow.

Once they see the ship or get abreast, probably hearing the engines and propeller much louder in that position, they sound and do not come back to the surface before the ship is 3 to 4 miles past. When their course crosses that of a ship the author has seen them dive under the keel to the other side in water as shallow as 20 fathoms. Near Cochin (India)

174

one of these whales migrating northwards was heading straight for a fisherman's nets, strung out in 12 fathoms of water. It approached the obstruction to about half a mile then carefully circumnavigated it.

In New Zealand waters 6 Sei Whales stranded between 1911 and 1968 all south of North Island and only the same number was observed alive during that period indicating that the species is rare in those waters. The Bryde's Whale of the next paragraph is more common and seems to be a permanent inhabitant of warmer regions northward of North Island.

In the middle of the South Pacific (N.E. of Ducie Isl.) a lone Sei Whale kept pace with the author's ship at a distance of 50 meters maintaining 19 knots for at least 10 minutes.

The name Sei was given by Norwegians because this whale always appeared on their coast at the same time as the Sei fish.

BRYDE's WHALE, plate nr. 93.

species	*Balaenoptera edeni* Anderson, 1878.
	[= *B. brydei* Olsen, 1912].
area	Tropical and subtropical seas, probably not in the Central Pacific Ocean (Chart 12).
sea temperature	15 to 30° C.
length	40 to 50 feet (12–15 m), females longer than males.
weight	± 18 tons (0.4 t/ft).
food	Fish (Herring, Mackerel and small Sharks to a length of 2 feet (60 cm). Once 15 Penguins were found in a stomach, no doubt by accident, but it brings the water temperature down to 15° C.
whalebone	270 pairs of plates, broad, thick and short. Maximum length 19″ (48 cm) with stiff long bristles. The colour in the front of the mouth is white, further back grey-black.
grooves	30 to 60 extending as far back as the navel, which is 3 to 4 feet longer than in Sei Whales.
illustration	From life and description. It is almost identical with the Sei Whale and only since 1912 recognized as a separate spec. On the underside the chin, throat and chest are dark where these are white in Sei. They also are very slender animals

compared to a more chubby Sei. The dorsal fin is reported to have a different shape; the author has not been able to verify in what way. From what he observed it is possibly more triangular.

special features	Indistinguishable from the Sei Whale in the field. Across the white belly (behind the dark chest) there is a grey band.
speed	Cruising 5 to 6 knots.
breath	Every 7 to 9 seconds.
schools	Single or perhaps in pairs.
biotope	Coastal and marine in the Atlantic and Indian Oceans.
migration	Unrecorded.
immatures	Calves are born throughout the year, length at birth is 14 feet (4.3 m) and they weigh 1 ton.
notes	The difference between this species and the Sei Whale as already mentioned cannot be noticed in the field.

The darker throat perhaps on rare occasions. The difference can serve if a caught or stranded animal must be identified. The author observed one in July 1966 only 3 miles West of Green Point (Cape Town) feeding continuously, swimming in small circles amongst well over a thousand penguins doing the same thing. It is therefore not surprising they occasionally swallow these birds. It paid no attention to the ship which had to alter course twice with wheel hard over to avoid ramming it.

Before 1961 about 5 Bryde's Whales were caught yearly off the N. coast of North Island (N. Zealand) where they seem to be in residence throughout the year without ever coming further south. Since the Whaling Station closed down their numbers are on the increase.

FIN WHALE or RAZORBACK, plate nr. 94.

species	*Balaenoptera physalus* (Linnaeus, 1758).
	[= *B.p. physalus* Northern hemisphere].
	[= *B.p. quoyi* Fischer, 1829, Southern hemisphere].
area	All oceans to – but not in – polar ice. Rare in the S.W. perhaps in the whole South Pacific (Chart 13).
sea temperature	0 to 30° C.
length	65 to 81 feet (20-25 m), females 1 to 2 feet longer than

	males. The northern race is $\pm$ 5 feet shorter than that of the south.
weight	To 64 tons (0.9 t/ft).
food	Krill and seasonal also small Fish (Herring, Capelin).
whalebone	265 to 470 pairs of plates. The left side is grey, on the right side $\frac{1}{3}$ in the front white, the rest grey. Fringe yellowish white. Length 20 to 36″ (50–90 cm).
grooves	68 to 114, 2 to 3″ (5–7.5 cm) apart extending backwards to beyond the navel.
illustrations	Life and photographs. The dorsal grey varies from light to dark. Underneath, including the undersurface of the flippers and flukes it is pure white. The head has asymmetric pigmentation. The left lower jaw on the outside the left baleens and the right lower jaw on the inside are gray. The right lower jaw on the outside, the right baleens and the left lower jaw on the inside are white.
special features	A long Finner comparable to the Blue Whale but much lighter in colour and with a lot of white. The Blue Whale has no white. After blowing which sometimes shows the snout, a very long back breaks the surface with at the rear end a fair sized dorsal fin (1 foot = 30 cm), followed by a rather short tailstock which has a sharp ridge. This ridge originated the name Razorback amongst whalers. The arching of the tailstock is specially noticeble when the animal swims fast. The flukes never show. They sometimes jump out of the water almost vertically, falling back sideways or belly upwards with an enormous splash.
speed	10 to 12 knots.
breath	5 or 6 times with shallow dives of 10 to 20 seconds then a deep dive lasting from 4 to 15 minutes with a maximum of 23 minutes. The blast is 15 to 20 feet (4.5–6 m).
schools	2 or 3, sometimes 100 and more, spread out.
biotope	Deep water. In arctic and antarctic waters during the summer staying outside the pack-ice.
migration	Migrating to the subtropics and sometimes the tropics during the winter. Their routes are about 200 miles off shore.
immatures	Pairing takes place from late autumn to early spring with a peak in winter. Calves are born after a little over 11

months, have a length of 22 feet (6.7 m), weigh ± 4 tons and are nursed for 6 months to a length of ± 39 feet (12 m). They reach maturity at 64 feet for males and 66 feet for females (19.5–20 m).

notes   This was and still is the most numerous member of the family. Since 1900 about 500.000 have been killed!

Their off shore habits keep them away from most of the regular shipping routes. The author has seen the breaching of this species north of Caroline Islands in the Pacific Ocean and southeast of Socotra Island in the Indian Ocean. Many were seen in June and July on the outside of the 100 fathom line between Cape Hatteras and New York.

Not more than 6 stranded in New Zealand waters between 1873 and 1968 and less than 10 were seen during the same period.

There seem to be more males than females, 55 %–45 %.

The blubber has a thickness of 2.5″ (6.5 cm).

BLUE WHALE, plate nr. 95.

species   *Balaenoptera musculus* (Linnaeus, 1758).

[= *B. m. musculus* Northern hemisphere].

[= *B.m. intermedia* Burmeister, 1866, Southern hemisphere].

[= *B.m. subspec.* (*pygmy*) Ichihara, 1961, Kerguelen].

area   All oceans, polar to subtropical, occasionally to tropical seas (Chart 14).

sea temperature   0 to 25° C.

length   90 to 100 and 110 feet (27.5–33.5 m), females 1 to 2 feet (30–60 cm) longer than males.

weight   To 150 tons (1.35 t/ft).

food   Krill. Stomach contents is one ton.

whalebone   270 to 395 pairs of plates, the colour including the fringe is black. They have a length of 23 to 41″ (59–104 cm) and a mouthful weights 220 lbs (100 kg).

grooves   70–118, 2 to 3″ (5–7.5 cm) apart, extending backwards to beyond the navel.

illustration   From life and photographs. The colour is slaty-blue all over, only the tips and the underside of the flippers are white. The blue is often mottled with paler and greyish patches and

178

	in the tropics the underside is sometimes yellowish with a film of diatoms. From this originates the name "sulphurbottom" amongst whalers.
special features	The enormous size and bluish mottled colour which the author judges leaning towards purplish. When the animal surfaces there appears after the blast a "long long" back which finally ends in a tiny dorsal fin. Of the short tailstock very little is seen and the flukes never show.
speed	10–12 knots, maximum 14 knots.
breath	12 to 14 shallow dives at 12 to 15 seconds intervals, then a deep dive lasting from 10 to 20 minutes reaching a max. of 50. The blast is 20 feet high (6 m).
schools	2 or 3 together, these small groups in greater formations of 30 to 50 widely spread out over an area 10 to 20 miles across.
biotope	During the summer in polar seas along and inside the pack-ice. Oceanic and marine. Deep water.
migration	To the polar areas in the spring and returning from there to warmer climates in autumn. Little is known about their whereabouts during winter. The migration routes are probably pelagic.
	Although they were very numerous in the Southern Oceans from South America eastwards to New Zealand only very few were seen along the east coast of South America, both coasts of South Africa and in Australian and New Zealand waters.
	Between 1873 and 1968 8 specimens stranded in New Zealand waters and 5 were taken by land stations during the same period.
	In comparison there were more in the Humboldt current along the South American west coast. Part of the North Pacific population probably migrates with Sei, Humpback and Sperm Whales to the Northern Indian Ocean on a route through the Moluccas and south of Java.
immatures	Pairing takes place over an extended period during the winter with a peak in June-July (south) and December-January (north). The calves are born after about 11 months usually one, occasionally two. They have a length of 24–25 feet (7.5 m), weigh 7.5 tons and are weaned after 7 months,

when ± 52 feet long (15.8 m).
Males reach maturity when 74 feet (22.5 m) long and females at 77 feet (23.5 m).

notes The widespread oceanic distribution makes observations few and far between, to which in the present day can be added that they are nearly extinct. In addition they are very shy and easily scared, sounding at the least unfamiliar noise. The author encountered them several southern winters well east of Socotra Island and also in the Gulf of Aden.

They have a rudimentary "moustache" of 4 hairs and a "beard" of 40 hairs. The weights of a 90 feet (27 m) female which had a total weight of 122 tons were as follows (figures in tons):

Meat	56.4	Blubber	25.7
Bones	22.6	Lungs	0.61
Tongue	3.2	Kidneys	0.55
Heart	0.63	Stomach	0.42
Liver	0.94	Largest vertebrae	0.24

The thickness of the blubber is about 4″ (10 cm).

The mothers milk is like "cows milk". Tits are 3.5″ (9 cm) these are concealed in two slits on either side of the genital opening. The male's penis has a length of 9.5 feet (2.9 m). Length of the flipper is 7.5 feet (2.3 m).

In 1892 one specimen reported as a Blue Whale washed up on a beach 18 miles south of Malacca (Malaya), in 1970 the author could identify the skull however as belonging to a Sei Whale.

genus   MEGAPTERA Gray, 1846.
species   HUMPBACK WHALE, plate nr. 96.
Megaptera novaeangliae (Borowski, 1781).
[= *M. nodosa* (Bonnaterre, 1789)].
[= *M.n. lalandi* (Fischer, 1829), Southern race].
area   All oceans (Chart 15).
sea temperature   0 to 30° C.
length   About 50 feet (± 17 m).
weight   To 45 tons (0.82 t/ft).
food   Krill, Shrimps and small Fish, sometimes Cod, once 6 Cormorants.

whalebone	250 to 350 pairs of plates, grey-black including the bristles, length to 24″ (60 cm).
grooves	18 to 26, 5 to 8″ apart (13 to 20 cm) extending backwards as far as the navel.
illustration	Life and photographs. They are grey-black above and dark below. Older animals have a white area between the flippers which extends with age. The body is covered with irregular knobs and growths, especially on the head and flippers. The flippers and the posterior edge of the flukes are largely white.
special features	A short thickset whale, the clown of the family in appearance as well as in behaviour. They are not afraid of ships and can be approached quite near especially when they are playing on the surface, jumping, rolling, splashing about and lobtailing. Their jumps are very spectacular, often repeatedly right out of the water and sometimes somersaulting 3 times backwards, waving about the enormous flippers (14 $\times$ 3.5 feet = 4.3 $\times$ 1.1 m). The splashes can be seen as far as the horizon. The flukes always show when sounding, which is always a vertical dive. The dorsal fin is small, more or less triangular of varied shape and thickness. The body is infested with parasites, whale lice, barnacles and even seaweed. The flukes have a spread of 27% of the total length.
breath	From 3 to 5 minutes on the surface, blowing every 15 to 20 sec. then a deep dive lasting from 3 minutes to a maximum of 20 min. The blast is said to be 5 to 6 feet high and broad. The author estimated from his own observation the blast to be at least 10 to 12 feet high.
speed	Cruising 3 to 4 knots, migrating perhaps to 6 knots considering the distances they cover.
biotope	Coastal but sometimes marine or oceanic, usually in deep water but also inside the 100 fathom line.
schools	Single or small groups of 3 to 4, bigger schools spread out over a wide area. The small groups several miles apart.
migration	During the summer in polar areas, in autumn migrating to subtropical and tropical latitudes, with evidence in all three oceans that they cross the equator. The travelling is usually

coastwise along fixed routes where in the course of time land-operated whaling stations were established which decimated their numbers. The whales who winter on the Makkran-coast (Persia-Pakistan) cover between 6000 and 7000 miles to and from the feeding grounds in the antarctic and between 8000 and 9000 miles to Kamchatka and the Bering Sea.

These must at least have a day's run of 100 miles. The author is under the impression that perhaps the majority migrates not further than the belt of 10–20° latitude and that under the stress of continuous pursuit they are becoming more pelagic, demonstrated by recent records of considerable schools of Humpbacks round the Hawaiian Islands and east of the Marqueses Islands.

immatures

Pairing takes place in late spring and is accompanied by much leaping and playing on the surface. The gestation period is 11–12 months. The calves are ± 14 feet (4.2 m) when they are born; one, sometimes two and weigh 1.1 ton. They are weaned after about 11 months at a length of 26 feet (8m) and reach maturity when 36–38 feet long (11–12m).

notes

It is a whale which ships frequently meet on the normal coastal shipping routes but was never very abundant in the north Atlantic. It has been so intensively hunted that it is now in danger of becoming extinct and is protected by the IWC by means of severe restrictions. It is also a known fact that these restrictions are not adhered to.

The decline of species in the SW Pacific is clearly demonstrated by the fact, that in autumn 1954: 495 of these whales migrated northwards through Cook Strait. In 1963 there were only 16, 9 of which were taken by the Whaling Station. Following spring only 11 passed on the way south.

These whales rarely strand and some remarkable facts have shown that they are not easily put out when trapped in shallow and confined places. One lived for 6 weeks in the river Tay (Scotland), another in Nantucket and Newport Harbour, once the home of the greatest whaling fleet of the world, and a third stayed one week in the Kuwait Inner Harbour (Persian Gulf) where it finally died after being hit by the propeller of a manoeuvring ship.

LITERATURE

Alpers, A., 1963: Dolphins.
    Butler and Tanner Ltd., Frome and London, pag. 1-251.
Anon, 1965: Stranding of 96 Pilot Whales near Yeu Island (France).
    Mammalia 29(1): 61-68.
Azzaroli, Maria L., 1968: Second specimen of *Mesoplodon pacificus*, the rarest living Beaked Whale.
    Monitore Zool. Ital. (N.S.) 2 (suppl.): 67-69.
Barnard, K. H., 1954: A guide to South African Whales and Dolphins.
    Guide no. 4 South Afr. Mus. Cape Town, 33 pp.
Bierman, W. H., and E. J. Slijper, 1947-48: Remarks upon the species of the genus *Lagenorhynchus* I and II.
    Kon. Ned. Akad. Wetensch. 50(10): 1343-1364; 51(1): 127-133.
Bree, P. J. H. van, 1966: On a skull of *Tursiops aduncus* (Ehrenberg, 1833) *Cetacea, Delphininae* found at Mossel Bay, South Africa in 1904.
    Ann. Natal Mus. 18(2): 425-427.
Bree, P. J. H. van, and J. Cadenat, 1968: On a skull of *Peponocephala electra* (Gray, 1846) (*Cetacea, Globicephalinae*) from Sénégal.
    Beaufortia (Amsterdam) 14(177): 193-201.
Bree, P. J. H. van, and R. Duguy, 1964: Sur un crâne de *Sotalia teuszi* (Kükenthal, 1892) (Cetacea, Delphinidae).
    Zts. f. Säugetierk. 30(5): 311-314.
Bree, P. J. H. van, and H. Nijssen, 1964: On three specimens of *Lagenorhynchus albirostris* (Gray, 1846) (Mammalia, Cetacea).
    Beaufortia (Amsterdam) 11(139): 85-93.
Bullis, H. R., and J. C. Moore, 1956: Two occurences of False Killer Whales and a summary of American records.
    Amer. Mus. Novitates, New York no. 1756, 5 pp.
Davies, J. L., 1960: The Southern Form of the Pilot Whale.
    J. Mammal. 41(1): 29-34.
Duguy, R., 1968: Note sur *Globicephala macrorhyncha* (Gray, 1846); un Cétacé nouveau pour les côtes de France.
    Mammalia 32(1): 113-117.
Duguy, R., and P. J. H. van Bree, 1968: Catalogue des Cétacés et des Pinnipèdes

du Musée d'Histoire Naturelle de la Rochelle.

Ann. de la Soc. des Sciences Nat. de la Charente Maritime, 4(9): 1-27.

Fraser, F. C., 1955: The Southern Right Whale Dolphin *Lissodelphis peroni* (Lacépède).

Bull. Brit. Mus. Nat. Hist. (Zool.) 2(11): 341-346.

Fraser, F. C., 1958: Common or Harbour Porpoises from French West Africa.

Bull. Inst. Français Afr. Noir. ser. A 1, 20: 276-285.

Fraser, F. C., and H. W. Parker, 1953: Guide for the identification and reporting of stranded whales, dolphins, porpoises and turtles on the British coasts.

Brit. Mus. Nat. Hist. London, 42 pp.

Gaskin, D. E., 1968: The New Zealand Cetacea.

Fish. Res. Bull. (New Series) no. 1, 92 pp.

Marine Dept. Fish. Res. Div., Wellington, New Zealand.

Gelder, R. G. van, 1960: Results of the Puritan-American Museum of Natural History Expedition to Western Mexico.

Marine Mammals from the coasts of Baja, California and the Très Marias Islands, Mexico.

American Mus. Novitates, New York no. 1992, 27 pp.

Herald, E. S., 1967: Bouto and Tookashee – Amazon Dolphins.

Pacific Discovery 20(1): 2-9.

Calif. Acad. of Sciences, San Francisco.

Kellogg, R., 1940: Whales, giants of the sea.

Nat. Geogr. Mag. 67: 35-90.

Mackintosh, N. A., 1965: The stocks of Whales.

Ed. Fishing New Books Ltd., London.

Mc.Cann, C., 1962 (a): The occurence of the Southern Bottle-nosed Whale, *Hyperoodon planifrons* Flower in New Zealand Waters.

Rec. Dominian Mus. Wellington N.Z. 4(3): 25-27.

Mc.Cann, C., 1962 (b): The taxonomic status of the Beaked Whale, *Mesoplodon hectori* (Gray) *Cetacea*.

Rec. Dominion Mus. Wellington N.Z. 4(9): 83-94.

Mc.Cann, C., 1962 (c): The taxonomic status of the Beaked Whale, *Mesoplodon pacificus* Longman, Cetacea.

Rec. Dominian Mus. Wellington N.Z. 4(10): 95-100.

Mc.Cann, C., 1963: Occurrence of Blainville's Beaked Whale in the Indian Ocean.

J. Bombay Nat. Hist. Soc. 60(3): 727-730.

Mc.Cann, C., 1964 (a): A further record of Blainville's Beaked Whale from the Indian Ocean.

J. Bombay Nat. Hist. Soc. 61(1): 179-180.

Mc.Cann, C., 1964 (b): The female reproductive organs of Layard's Beaked Whale, *Mesoplodon layardi* (Gray).

Rec. Dominian Mus. Wellington N.Z. 4(23): 311-316.

Mc.Cann, C., and P. H. Talbot, 1963: The occurrence of True's Beaked Whale, *Mesoplodon mirus True*, in South African Waters, with a key to South African species of the genus.
Proc. Linn. Soc. London, 175(2): 137-144.

McLachlan, G. R., R. Liversidge and R. M. Tietz, 1966: A record of *Berardius arnouxi* from the South-East coast of South Africa.
Ann. of Cape Prov. Mus. 5: 91-109.

Melville, H., 1961: Moby Dick.
Ed. Wash. Square Press, New York, 402 pp.

Mohr, E., 1964: Notizen über den Flussdelphinen *Inia geoffrensis* de Blainville, 1817.
Der Zool. Garten N.F. 29(5): 262-270.

Moore, J. C., 1963: Recognizing certain species of Beaked Whales of the Pacific Ocean.
Am. Midl. Nat. 70(2): 396-428.

Moore, J. C., 1968: Relationships among the living genera of Beaked Whales, with classification, diagnoses and keys.
Fieldiana: Zoology 53(4): 209-298.
Field Mus. Nat. Hist., Chicago.

Moore, J. C., and R. S. Palmer, 1955: More Piked Whales from the Southern North Atlantic.
J. Mamm. 36(3): 262-433.

Moore, J. C., and F. G. Wood Jr., 1957: Differences between the Beaked Whales, *Mesoplodon mirus* and *Mesoplodon gervaisi*.
Amer. Mus. Novitates no. 1831: 1-25.

Mörzer Bruyns, W F. J., 1960: The Ridge-backed Dolphin of the Indian Ocean.
Malayan Nature J. 14: 159-165.

Mörzer Bruyns, W. F. J., 1966: Some notes on the Irrawaddi Dolphin, *Orcaella brevirostris* (Owen, 1866).
Zts. f. Säugetierk. 31(5): 367-370.

Mörzer Bruyns, W. F. J., 1968: Sight records of Cetacea belonging to the genus *Mesoplodon*.
Zts. f. Säugetierk. 33(2): 106-107.

Mörzer Bruyns, W. F. J., 1969: Notes on the Hector's Dolphin, *Cephalorhynchus hectori* of New Zealand.
Rec. Dominion Mus. Wellington (N.Z.) (in the press).

Mörzer Bruyns, W. F. J., 1969: Sight records and notes on the False Killer Whale *Pseudorca crassidens* (Owen, 1846).
Säugetierkundliche Mitteilungen 17(A): 351-356.

Nakajima, M., and M. Nishiwaki, 1965: First occurrence of a porpoise *Electra electra* in Japan.
Sci. Repts Whales Res. Inst. (Tokyo) 19: 91-104.

Nishimura, S., and M. Nishiwaki, 1964: Records of the Beaked Whale, *Mesoplodon* from the Japan Sea.
Publ. Seto Mar. Biol. Lab. 12(4): 323-334.

Nishiwaki, M., 1965: *Feresa attenuata* captured at the Pacific coast of Japan.
Sci. Rep. Whales Res. Inst. (Tokyo) 19: 65-90.

Nishiwaki, M., 1966: Distribution and migration of Marine Mammals of the North Pacific Area. Ocean Res. Inst. Univ. of Tokyo.
11th Pac. Science Congress Symp. no. 4: 1-49.

Norman, J. R., and F. C. Fraser, 1948: Giant Fishes, Whales and Dolphins.
Ed. Putman, London, 361 pp.

Norris, K. S., 1961: Standardized methods for measuring and recording data on the smaller Cetaceans.
J. Mamm. 42(4): 471-476.

Norris, K. S., and J. H. Prescott, 1961: Observations on Pacific Cetaceans in Californian and Mexican Waters.
Univ. Calif. Publ. Zool. 63(4): 291-402.

Oliver, W. R. B., 1937: *Tamacetus shepherdi*, a new genus and species of Beaked Whales from New Zealand.
Proc. Zool. Soc. London (B), 1937(3): 371-381.

Oliver, W. R. B., 1946: A pied variety of the coastal Porpoise.
Dom. Mus. Rec. Zool. 1(1): 1-4.

Omura, H., K. Fujino and F. Kimura, 1955: Beaked Whale, *Berardius bairdi* of Japan, with notes on *Ziphius cavirostris.*
Sci. Rept. Whales Res. Inst. (Tokyo) 10: 89-132.

Omura, H., 1958: North Pacific Right Whale.
Sci. Repts. Whale Res. Inst. (Tokyo) 13: 1-52.

Omura, H., and H. Sakiura, 1956: Studies on the Little Piked Whale from the coast of Japan.
Sci. Repts. Whale Res. Inst. (Tokyo) 11: 1-37.

Orr, R. T., 1948: A second record of Cuvier's Whale from the Pacific Coast of the United States.
J. Mamm. 29(4): 420-421.

Orr, R. T., 1953: Beaked Whales (*Mesoplodon*) from California with comments on taxonomy.
J. Mamm. 34(2): 239-249.

Sanderson, J. T., 1956: Follow the Whale.
Boston Ed. Little, Brown and Co. Ltd., 423 pp.

Scheffer, B. V., and D. W. Rice, 1963: A List of the Marine Mammals of the World.
U.S. Fish- and Wildlife Serv. Spec. Sc. Rep. Fisheries no. 431, Washington, 12 pp.

Slijper, E. J., 1962: Whales.
New York, 475 pp.

Slijper, E. J., 1964: Hundred years after Svend Foyn.

Vakblad voor Biologen (Amsterdam) 44: 199-211.

Slijper, E. J., W. L. van Utrecht and C. Naaktgeboren, 1964: Remarks on the distribution and migration of Whales based on observations from Netherlands ships.

Bijdr. tot de Dierk. (Amsterdam) 34: 93 pp.

Utrecht, W. L. van, and S. v. d. Spoel, 1961: Observations on a Minke Whale. (*Mammalia Cetacea*) from the Antarctic.

Zts. f. Säugetierk. 27(4): 217-221.

Yamada, M., 1954: An account of a rare porpoise, *Feresa* Gray from Japan.

Sci. Repts. Whale Res. Inst. (Tokyo) 9: 59-88.

TABLE TO ASSIST RECOGNITION OF DOLPHINS AND WHALES AT SEA
BY COMPARING THE SIZE OF THE DORSAL FINS AND ESTIMATING
THE LENGTH OF THE BEAK

For stranded specimens the spread of the flukes for some species is added

The first thing an observer usually sees of a dolphin at sea, is the dorsal fin, which invariably shows above the surface when the animal blows, next the outline of the head and the size of the beak follow later.
The dorsal fin is judged as part of and relation to the body more often than by actual size.
The 9″ fin of a Common Dolphin is "high" whilst the 12″ fin of the Fin Whale of 70 feet is "very low".
Beaks are observed and judged on their own and in relation to the head by means of binoculars. In the field the actual size is more important than the percentage of the total length.
For some ill-defined beaks as in the genus Mesoplodon, the length is roughly half the distance from the tip of the snout to the blowhole.
The measurements in the table are taken as much as possible from zoological publications; where these were not available good photographs of the animals concerned, taken in profile, were used.
The first three columns give the name of the dolphin; sex (if the sexes have different sizes), age, "a" (adult), "i" (immature) and the total length of the animal.
The next three columns give the size of the dorsal fin in cm., inches and as % of total body length, followed by the same particulars of the beaks.
The last two columns give the spread of the flukes in cm. and in % of total length.
Although in all branches of science the metric system offers a more accurate and uniform way of registering data which have to be compared with each other, the author prefers feet and inches in the field.
The larger "unit" and the relationship to the human body make estimation easier, especially as the majority of the dolphins are roughly our own size.
The sequence in the table is that of the dorsal fins from high to low, 8 % can be considered a "normal height".

TABLE FOR COMPARING DORSAL FINS AND BEAKS OF SOME SPECIES

Whale or Dolphin (Wh.) (D.)	sex age	lth. m	height d. fin cm	d. fin inch	%	length of beak cm	beak inch	%	spr. flukes cm	flukes %	ref. nr.
I: Very high more than 10 % of total length											
Killer	m	8.00	180	6'00"	22.5	—	—	—	200	25	69
id.	f	—	140	4'06"	17.0	—	—	—	150	—	—
Grampus	—	4.00	44	1'05"	11.0	—	—	—	—	—	64
Gulfstr. Spot. D.	m	2.15	24	10"	11.0	13	5.2"	6.0	53,5	24,5	41
Pygmy Killer	m	2.30	24	10"	10.5	—	—	—	62	27	68
id.	f	2.18	22	9"	10.2	—	—	—	63	27	68
II: High 8–10 % of total length											
Little Killer	—	2.60	25	10"	9.6	—	—	—	65	25	67
False Killer	m	6.10	56	1'09"	9.3	—	—	—	—	25	66
id.	f	4.60	43	1'05"	9.3	—	—	—	—	25	66
Baird's D.	m	1.69	16	6.4"	9.5	15	6"	8.0	35	20,5	83
Black White D.)	f	1.65	16	6.4"	9.5	15	6"	8.0	35	20,5	28
Pac. White Sid. D.	m	2.28	21,5	8.7"	9.5	2	1"	1.0	56	25	21
id.	f	2.04	19,5	7.8"	9.5	2	1"	1.0	51	25	21
Common D.	—	2.20	22	10.0"	10.0	16	6.0"	7.3	55	23	25
Long beaked D.	—	2.00	17	7"	8.5	21	8.4"	8.5	46	23	30 31
Euphrosyne D.	—	2.16	18	7.2"	8.8	11	4.4"	5.1	46	21,5	37 39
Pilot Wh.	m	5.50	44	1'05.5"	8.0	—	—	—	132	24	65
id.	i	1.83	18	7"	9.7	—	—	—	49	27	65
Dall's Porp.	m	2.00	17	7"	8.5	—	—	—	50	25	12
Bottle-nosed D.	—	3.00	25	10"	8.2	10	4"	3.3	70,5	23,5	60
Atlantic Sp. D.	—	2.00	16	6"	8.0	14,5	6"	7.4	44	22	43
III: Medium 6–8 % of total length											
Guyana River D.	—	1.90	13,5	5.4"	7.4	12	5"	7.0	—	—	54
Hector's D.	—	1.50	10,5	4"	7.0	—	—	—	46	31	17
Pilot Wh.	f	4.20	27	11"	6.5	—	—	—	88	21	65
id.	i	2.08	15	5"	7.3	—	—	—	46	22	65
G.o. Panama Sp. D.	—	2.24	15	6"	6.7	16,2	6"	7.2	47	20,5	42
Cuvier's Wh.	—	6.40	41,5	1'04.5"	6.5	45	1'04.5"	7.0	79	23	81
Common Porpoise	—	1.40	9	3.5"	6.3	—	—	—	33	23,5	7
V: Low 4–6 % of total length											
Leadcoloured D.	—	1.80	10	4"	5.5	14,5	6"	8.0	—	—	56
N. Bottle-nosed Wh.	m	10.00	46	1'6"	4.6	43	1'00"	4.3	—	—	84
. Bottle-nosed Wh.	m	6.36	29	11.5"	4.6	21,5	8.5"	3.4	164	26	84
Sabre-toothed Wh.	—	5.17	21,5	8.5"	4.2	29	11.5"	5.6	128	25	77a
id.	i	2.39	12	5"	5.0	—	—	—	61	26	77a
Ginkgo Wh.	—	5.20	26	10"	5.0	27	10.5"	5.2	1.57	30	78

Whale or Dolphin (Wh.)	sex age	lth. m	height d. fin cm	inch	%	length of beak cm	inch	%	spr. flukes cm	%	ref nr
IV: continued:											
True's Wh.	—	5.10	25,5	10"	5.0	21	8"	4	1.27	25	75
Gulfstream Wh.	—	4.80	19	7.7"	4.0	19	7.5"	4	1.44	30	74
Blainville's Wh.	—	4.40	—	—	± 4.0	20	8"	4.5	—	—	80
Straptoothed Wh.	—	4.90	—	—	± 4.0	31	12"	6.2	—	—	79
Scamperdown Wh.	—	4.00	—	—	± 4.0	22,5	9"	± 5	1.10	27	76
Tasman's Wh.	—	6.00	—	—	?	45	18"	9.4	—	—	72
Minke Wh.	f.i.	2.80	12,7	5"	4.5	—	—	—	—	22	91
id.	m	8.20	34	1'01.5"	4.0	—	—	—	2.54	30	91
V: Very low, less than 4 % total length											
Baird's Wh.	m	10.64	26,6	10.5"	2.5	65	2'02"	6.2	2.93	28	83
id.	f	10.94	30,3	12"	2.8	—	—	—	3.06	28	83
Arnoux' Wh.	—	9.30	14,9	6"	1.6	± 54	1'09"	5.5	2.65	27	82
Amazone River D.	m	2.70 ±	5,5	2"	2	± 24	9.5"	12.6 ±	0.46	17	53
Sperm Wh.	m	20.00 ±	30	12"	1.5	—	—	—	5.00	25	85
Pygmy Sperm Wh.	—	4.00	16	6.5"	4	—	—	—	—	26	86
Pygmy Right Wh.	—	6.00 ±	12	5"	2	—	—	—	—	± 20	89
Irrawaddi D.	—	2.10	4	1.5"	2	—	—	—	—	± 28	71
Gray Wh.	a	13.00	—	—	—	—	—	—	—	—	90
id.	i	3.34	12,5	5"	3.5	—	—	—	—	—	90
Sei Wh.	—	18.00 ±	64	2'01"	3.5	—	—	—	—	± 28	92
Fin Wh.	—	22.00 ±	36	1'02"	1.5	—	—	—	—	± 28	94
Blue Wh.	—	30.00 ±	30	12"	1	—	—	—	—	± 28	95
Humpback Wh.	—	15.00 ±	30	12"	2	—	—	—	4.05	27	96
VI: No dorsal fin											
Greenland Wh.	—	18.00	—	—	—	—	—	—	8.00	43	88
Black Right Wh.	—	17.00	—	—	—	—	—	—	7.00	41	87
Peron's D.	—	1.80	—	—	—	—	—	—	—	—	50
Peale's D.	—	2.18	—	—	—	7,5	3"	3.5	0.37	17	87
Black Finless Porp.	—	1.50	—	—	—	—	—	—	—	—	1

N.B.: The ± figures are estimates

TABLE TO HELP IDENTIFICATION OF DOLPHINS AND WHALES BY THE AVERAGE NUMBER OF TEETH OR BALEENS – IN EACH HALF JAW

Latin name	ref. nr.	upper	lower	common name
Monodon (upper jaw only)	6	1	0	Narwhal

Lower jaws only Beaked Whales and Sperm Whales

Latin name	ref. nr.	upper	lower	common name
Mesoplodon	73/80	0	1	Beaked Whales
Ziphius	81	0	1	Cuvier's Whale
Hyperoodon	84	0	1	Bottlenosed Whale
Berardius	82/83	0	2	Duvernoy's Whales
Physeter	85	0	24/30	Sperm Whale
Kogia	86	0	9/15	Pygmy Sperm Whale

Less than ± 15 in each half jaw (Killer Group)

Latin name	ref. nr.	upper	lower	common name
Delphinapterus	5	10	8	Beluga
Grampus	64	0/2	.3/7	Risso's Dolphin
Globicephala	65	10	10	Pilot Whale
Pseudorca	66	8/11	8/11	False Killer
Feresa	68	10/11	11/13	Pygmy Killer
Orcaella	71	14/17	14/15	Irrawaddi Dolphin
Orcinus	69	13/12	12/12	Killer Whale

± 15 to 25 in each half jaw (a.o. Porpoises)

Latin name	ref. nr.	upper	lower	common name
Neophocaena	11	15/19	15/19	Black Finless Porpoise
Phocoena spinnipinnis	10	16/17	16/17	Bürmeister's Porpoise
Phocoena dioptrica	9	19/21	19/21	Spectacled Porpoise
Phocoena phocoena	7	22/27	21/25	Common Porpoise
Phocoena sinus	8	—	—	Gulf of California Porpoise
Phocoenoides truei	13	19	22	True's Porpoise
Phocoenoides dalli	12	27/25	25/30	Dall's Porpoise
Tursiops	60/62	20/25	20/25	Bottlenosed Dolphins
Tasmacetus	72	19	26	Tasman's Beaked Whale
Steno	51	20/27	20/27	Rough-toothed Dolphin

± 25 to 32 in each half jaw (Southern Dolphins, River/Estuary Dolphins)

Latin name	ref. nr.	upper	lower	common name
Peponocephala	67	24	24/25	Little Killer
Tursiops aduncus	63	26	27	Red Sea Dolphin
Cephalorhynchus heavisidei	16	25/30	25/30	Heaviside Dolphin
Cephalorhynchus commersoni	14	29/30	29/30	Commerson's Dolphin
Cephalorhynchus hectori	17/18	24/28	26/28	Hector's Dolphin
Cephalorhynchus eutropia	15	29/30	29/31	Chilian (White Bellied) Dolphin
Lagenorhynchus albirostris	19	25/26	25/26	White Beaked Dolphin
Lagenorhynchus acutus	20	30/34	30/34	White-sided Dolphin
Lagenorhynchus obliquidens	21	29/31	30/32	Pac. White-sided Dolphin
Lagenorhynchus cruciger	22	28	28	Hourglass Dolphin
Lagenorhynchus obscurus	23	30/36	30/36	Dusky Dolphin
Platanista gangetica	1	29	29	Ganges Dolphin
Inia geoffrensis	2	25/28	25/28	Amazone Dolphin or Boto
Sotalia pallida	53	28/31	28/31	Amazone River Dolphin

TABLE OF TEETH AND BALEENS CONTINUED

latin name	ref. nr.	upper	lower	common name
Sotalia guianensis	54	30	30	Guiana River Dolphin
Sousa teuszi	55	29/30	27	Cameroun River Dolphin
Sousa chinensis	59	32	32	Chinese White Dolphin
± 33 to 39 in each half jaw (a.o. River and Estuary Dolphins)				
Lipotes	3	33/36	33/36	Chinese Lake Dolphin
Sousa plumbea	56	34/38	34/38	Leadcoloured Dolphin
Sousa lentiginosa	57	36	36	Freckled Dolphin
Sousa borneensis	58	37/39	37/39	Borneo White Dolphin
Delphinidae spec.	27	35/38	35/38	Agulhas Dolphin
Stenella plagiodon	41	34/37	34/37	Gulfstream Spotted Dolphin
Stenella frontalis	43	37/38	37/38	Atlantic Spotted Dolphin
Stenella malayana	45	39	39	Malayan Dolphin
± 40 to 50 in each half jaw (Oceanic Dolphins)				
Stenella attenuata	44	35/44	35/44	Philippines Dolphin
Stenella graffmani	42	43	40	G. of Panama Spotted Dolphin
Stenella euphrosyne	37/39	40/46	40/46	Euphrosyne Dolphin
Stenella caeruleoalba	36	50	50	Blue White Dolphin
Stenella or Delph. roseiventris	29	48	45	Red Bellied Dolphin
Delphinus delphis (capensis)	25/26	40/50	40/50	Common Dolphins
Lissodelphis borealis	49	44	47	Northern Right Whale Dolphin
Lissodelphis peroni	50	43	43	Southern Right Whale Dolphin
More than 50 in each half jaw				
Pontoporia	4	50/60	50/60	La Plata Dolphin
Stenella longirostris/microps	30/35	53/55	51/52	Long Beaked Dolphins

Whalebone Whales in upper jaws only

	ref. nr.	upper	length	throat folds	
Balaena glacialis	87	250	9'0"	0	Black Right Whale
Balaena mysticetus	88	300/360	13/15'	0	Greenland Whale
Caperea marginata	89	230	2'3"	0	Pygmy Right Whale
Eschrichtius	90	150	1'4"	2/4	Gray Whale
Balaenoptera acutorostratus	91				Minke Whale
Balaenoptera N. Atlantic		270/348	12"	50/60	
Balaenoptera N. Pacific		230/295	12"		
Balaenoptera borealis	92	340	26"	32/62	Sei Whale
Balaenoptera edeni	93	270	19"	30/60	Bryde's Whale
Balaenoptera physalus	94	265/340	30/36'	68/114	Fin Whale
Balaenoptera musculus	95	270/395	23/41'	70/118	Blue Whale
Megaptera novae angliae	96	250/350	24"	18/26	Humpback Whale

Unknown number of teeth

From a number of dolphins mentioned in this guide the number of teeth is not known.
These are: Serawak D., South China Sea D. (24), Malacca D. (24a), Black White D. (28),
Java Sea D. (29a), Greek D. (40), Flores Sea D. (46), Senegal D. (47), Illigan D. (48), Elliot's D.
(52) and the Alula Killer Whale (70).

WHAT TO RECORD AND MEASURE OF A STRANDED OR CAPTURED WHALE OR DOLPHIN

(Standardized method: The Committee on Marine Mammals
American society of Mammologists)

with some additions by the author to include: beak, melon and scars

DATA
a. species
b. number
c. sex and age
g. locality
h. condition of specimen(s)
i. remarks, circumstances of stranding or capture
j. weight
k. number of teeth or baleens:
   right upper . . . . . .
   right lower . . . . . .
   left upper . . . . . .
   left lower . . . . . .
l. dimensions largest and smallest tooth: height . . . . length . . . . width . . . .
   if only one or two pairs of teeth in lower jaw:
   distance tip lower jaw to the anterior margin (each) pair
   distance across (each) pair
m. number of throat grooves
n. colour pattern (sketch)
d. date of stranding or capture
e. time of stranding or capture
f. observer

MEASUREMENTS (body) all along straight lines parallel to axis body
1 Total length: "Tip" of upper jaw to deepest part notch in flukes
2 Tip to centre of the eye
3 Tip to apex of melon (is length of beak)
3a Height and width of beak at tip
3b Height and width of beak in front of melon
3c Height and width 3" behind apex melon
4 Tip to angle of the gape (corner of the mouth)
5 Tip to earopening

6   Centre eye to earopening (direct)
7   Centre eye to angle of the gape (direct)
8   Centre eye to centre blowhole (direct)
9   Tip to centre blowhole (s)
10   Tip to anterior insertion of flipper
11   Tip to anterior insertion of dorsal fin
11a Tip to posterior edge of dorsal fin
12   Tip to centre navel
13   Tip to centre genital aperture
14   Tip to centre anus
15   Projection lower jaw beyond (or behind) upper jaw
16   Tip to posterior end of throat grooves
17   Thickness blubber at anterior insertion of dorsal fin
18   Thickness blubber middle flanks at midlength
19   Thickness blubber mid-ventral at midlength
20   Length throatgrooves: maximum and minimum
21   Girth (if possible take diameters) behind flippers
21a Girth over blowhole
22   Girth maximum, state distance from tip
23   Girth at anus
23a Height and width tailstock in front of flukes
24   Dimensions eye: height and length . . . .
25   Length mammary slits: right . . . . left . . . .
26   Length genital slit . . . . and anal opening . . .
27   Dimensions blowhole(s) width . . . . length . . . .
28   Diameter earopening right . . . . left . . . .
29   Flipper: length, anterior insertion to tip
30   Flipper: length axila (posterior insertion) to tip
31   Flipper: width maximum
32   Dorsal fin: height . . . . width of point one inch from tip . . . .
33   Dorsal fin: length of base
34   Flukes: width tip to tip . . . . Width of points one inch from tip
35   Flukes: length in the middle to posterior margin
36   Flukes: depth of notch

LIFE AND HISTORICAL DATA:
Scars: length and width of prominent scars
            distance between parallel scars
            percentage of surface covered by scars on head, back, flanks, underside,
            tailstock, flukes
External parasites:
and (for zoologists):

Internal parasites:
Nasal passages
Intestines
Air sinusses
Blubber
Stomach
Kidney
Others:

Stomach contents: (preserve in 10 % formaline if possible)
Reproduction:
female ovaries
follicles, largest diameter left . . . . right . . . .
active corpora lutea (no) left . . . . right . . . .
scars of corpora albicantra (no) left . . . . right . . . .
foetus length . . . . sex . . . .
milk in mammary glands: much . . . . some . . . . none . . . .
male testes
dimension left . . . . x . . . right . . . . x . . .
weight left . . . . right . . . .
sperm in cut epididymus present . . . . absent . . . .
Preserve in 10 % formaline: both ovaries, any small foetus, 1 cm³ of mammary glands and testes

ALPHABETICAL INDEX

GANGES DOLPHIN 1 life + illust.

WMB.

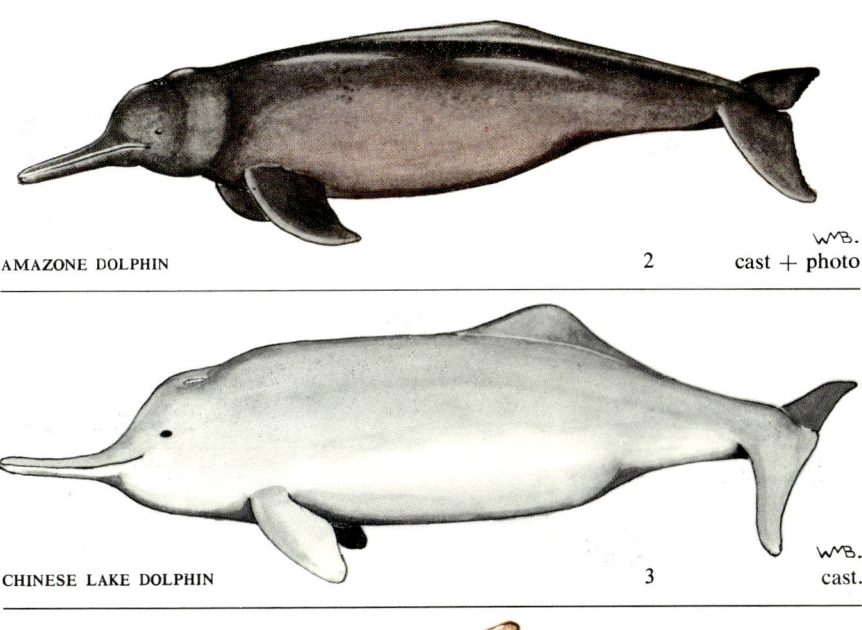

AMAZONE DOLPHIN 2 cast + photo

WMB.

CHINESE LAKE DOLPHIN 3 cast.

WMB.

LA PLATA DOLPHIN 4 cast + photo

WMB.

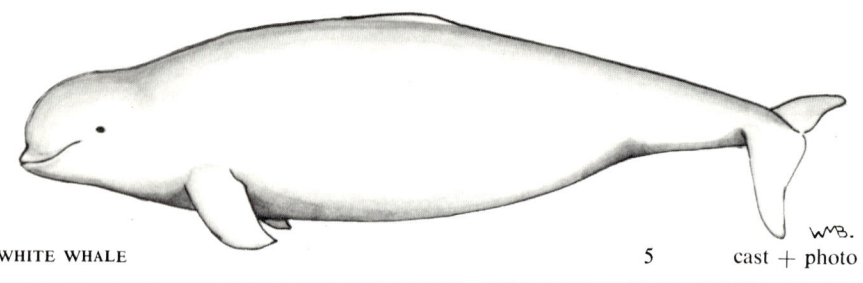

WHITE WHALE   5   cast + photo

NARWHAL   6   illustr

COMMON PORPOISE   7   life + cast

GULF OF CALIFORNIA PORPOISE   8   description

SPECTACLED PORPOISE   9   description

BURMEISTER PORPOISE   10   description

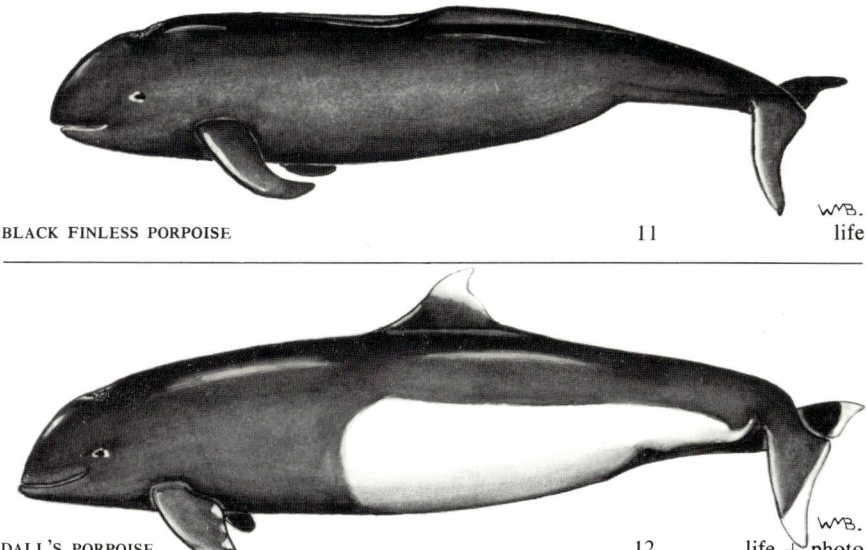

BLACK FINLESS PORPOISE   11   life

DALL'S PORPOISE   12   life + photo

TRUE'S PORPOISE   13   description

COMMERSON'S DOLPHIN   14   description

WHITE BELLIED DOLPHIN   15   description

HEAVISIDE DOLPHIN   16   description

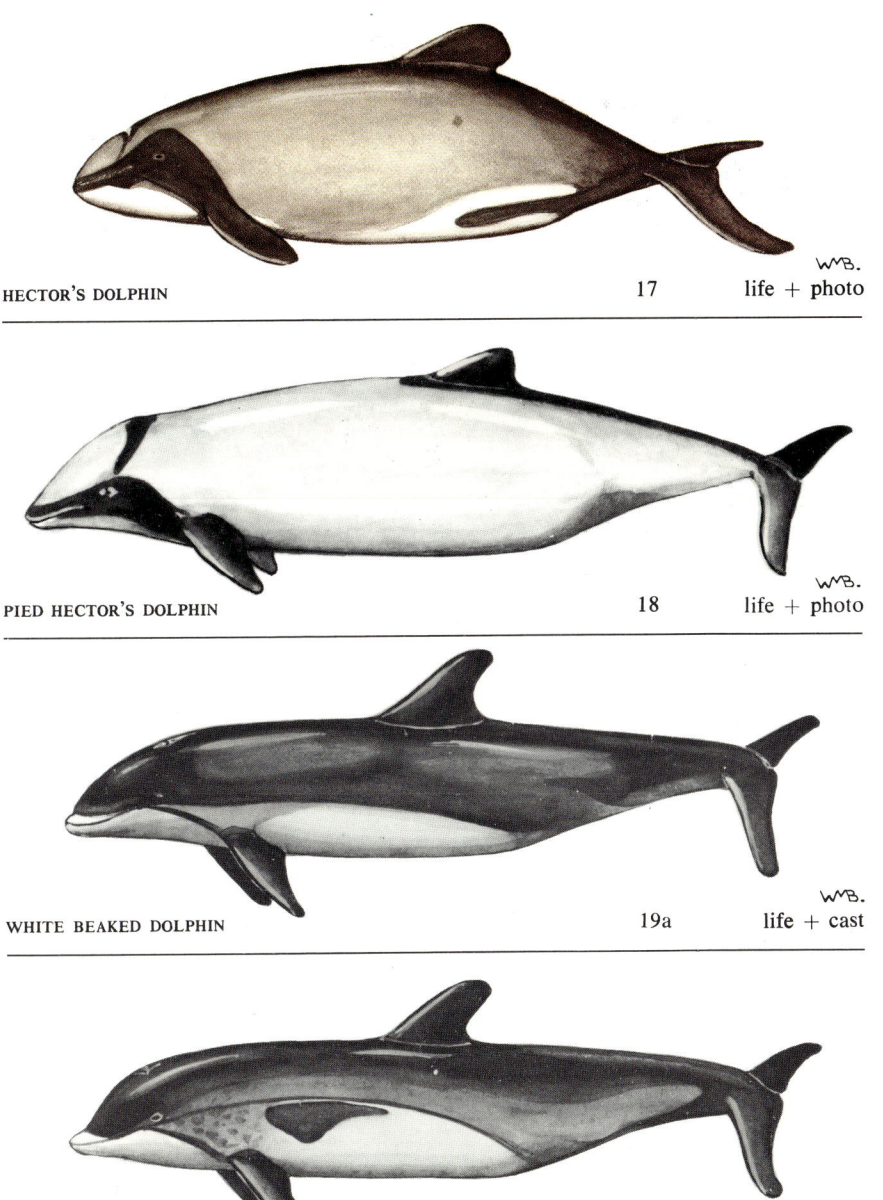

HECTOR'S DOLPHIN   17   life + photo

PIED HECTOR'S DOLPHIN   18   life + photo

WHITE BEAKED DOLPHIN   19a   life + cast

WHITE BEAKED DOLPHIN   19b   life + cast

WHITE SIDED DOLPHIN   20 life + cast + photo   W͜B.

PACIFIC WHITE SIDED DOLPHIN   21a   life + photo   W͜B.

PACIFIC WHITE SIDED DOLPHIN   21b   life + photo   W͜B.

HOURGLASS DOLPHIN   22   cast + photo   W͜B.

DUSKY DOLPHIN

23a-b-c life + cast + photo

WMB.

DUSKY DOLPHIN

23d life + cast + photo

WMB.

STH. CHINA SEA DOLPHIN

24

life

WMB.

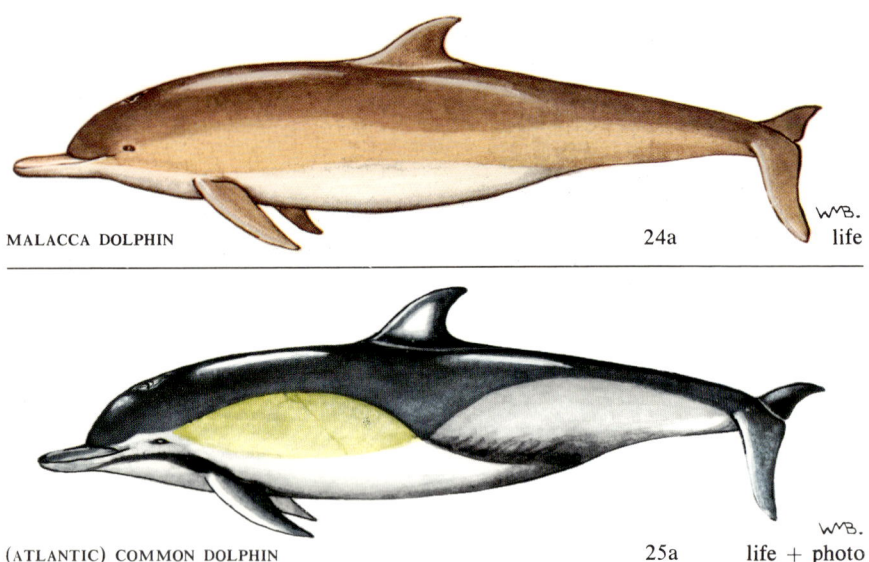

MALACCA DOLPHIN   24a   life

W͡B.

(ATLANTIC) COMMON DOLPHIN   25a   life + photo

W͡B.

(EAST MED.) COMMON DOLPHIN   25b   life

W͡B.

CAPE DOLPHIN   25c   life

W͡B.

(PACIFIC) COMMON DOLPHIN   26   life + photo

COMMON DOLPHIN   26a   life

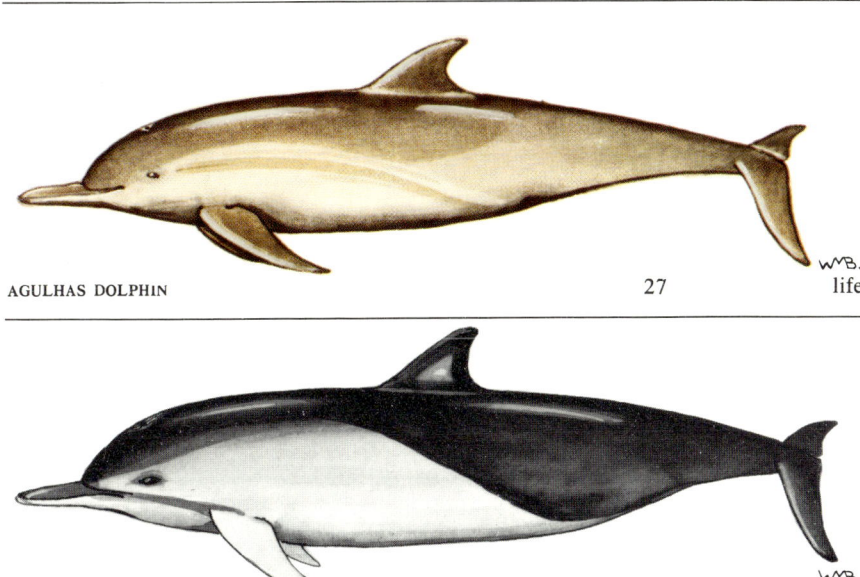

AGULHAS DOLPHIN   27   life

BLACK WHITE DOLPHIN   28   photo + life

RED BELLIED DOLPHIN   29   W.B.
                                                              life

JAVA SEA DOLPHIN   29a   W.B.
                                                              life

LONG BEAKED DOLPHIN   30   W.B.
                                                              life

LONG BEAKED DOLPHIN   31   W.B.
                                                              life

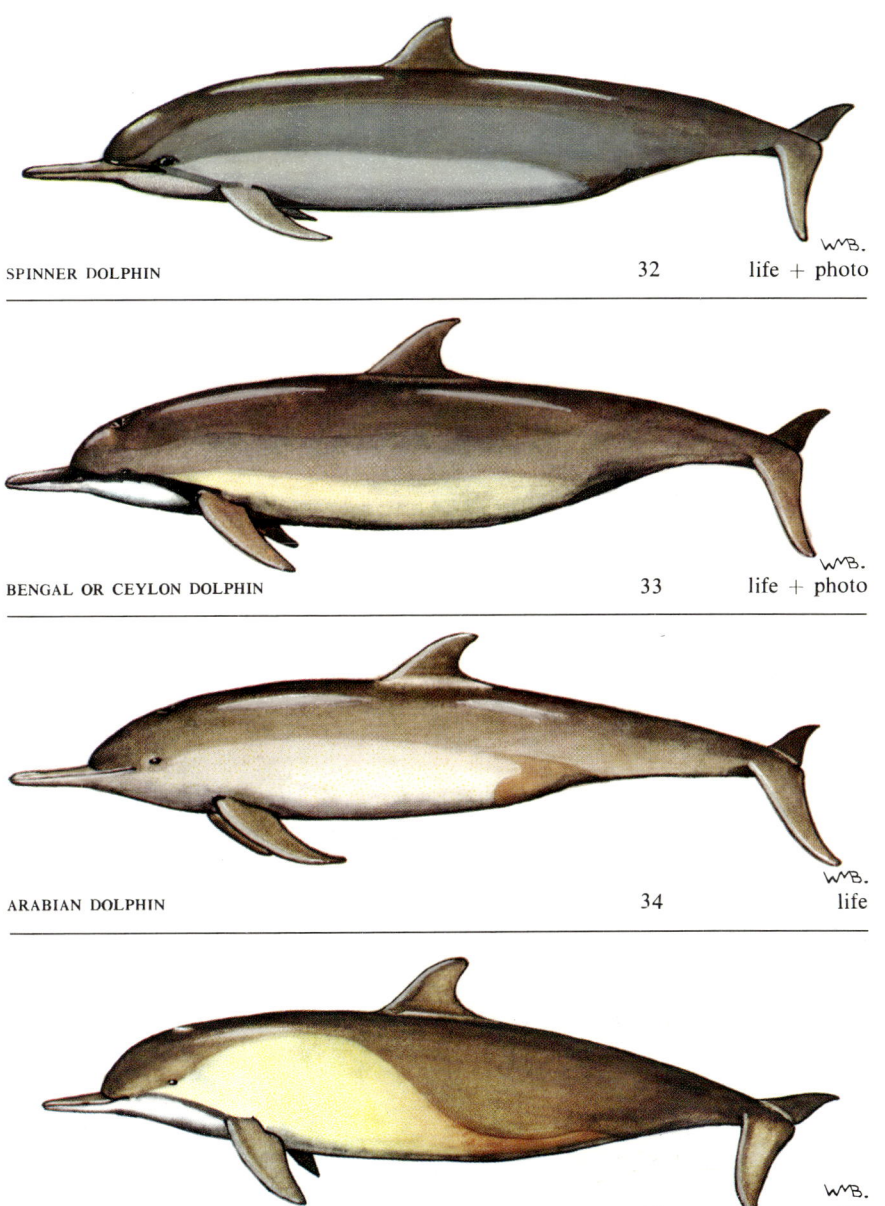

SPINNER DOLPHIN   32   life + photo

BENGAL OR CEYLON DOLPHIN   33   life + photo

ARABIAN DOLPHIN   34   life

GALAPAGOS DOLPHIN   35   life

BLUEWHITE DOLPHIN   36   WMB.
cast + descr.

EUPHROSYNE DOLPHIN   37   WMB.
life

EUPHROSYNE DOLPHIN   38   WMB.
life + photo

EUPHROSYNE DOLPHIN   39   WMB.
life

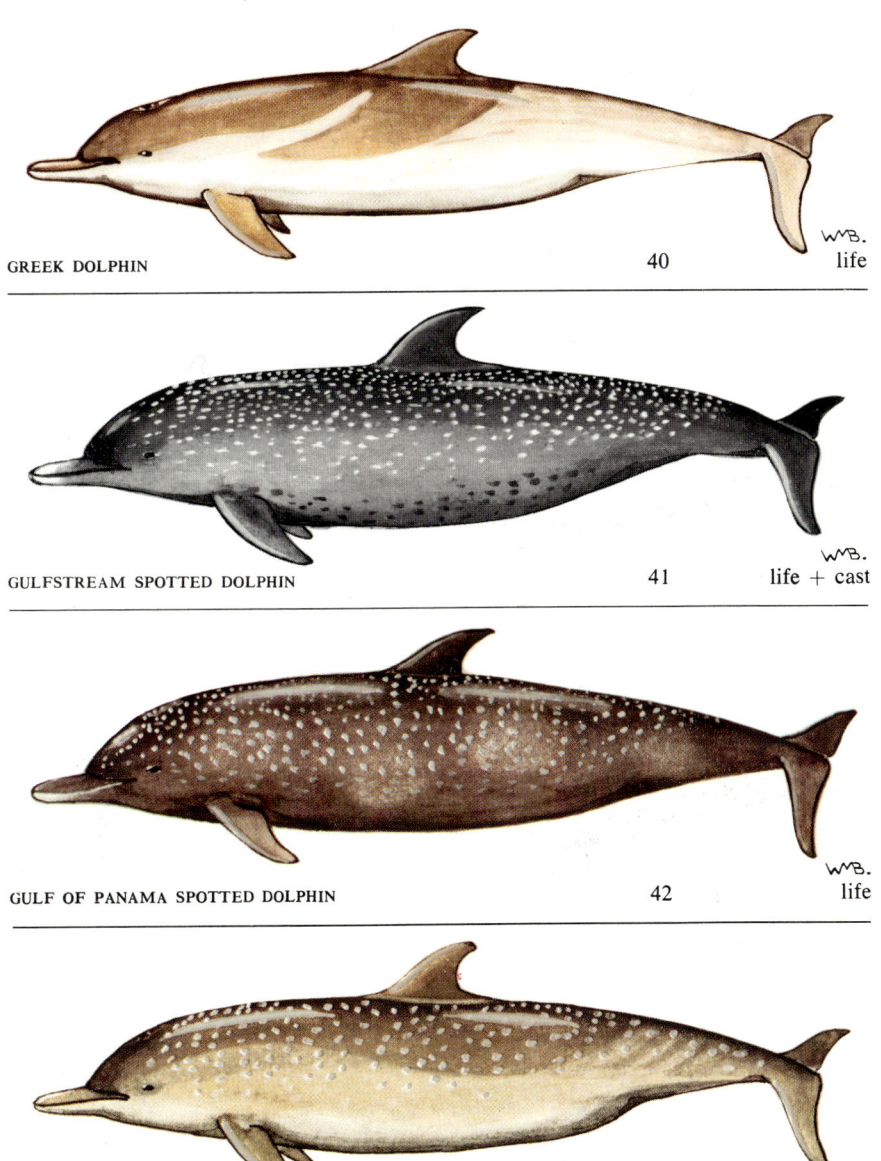

GREEK DOLPHIN 40 life

GULFSTREAM SPOTTED DOLPHIN 41 life + cast

GULF OF PANAMA SPOTTED DOLPHIN 42 life

ATLANTIC SPOTTED DOLPHIN 43 life

PHILIPPINE DOLPHIN   44   WMB. life

MALAY DOLPHIN   45   WMB. life

FLORES SEA DOLPHIN   46   WMB. life

SENEGAL DOLPHIN   47   WMB. life

ILLIGAN DOLPHIN   48   WMB.
                                              life

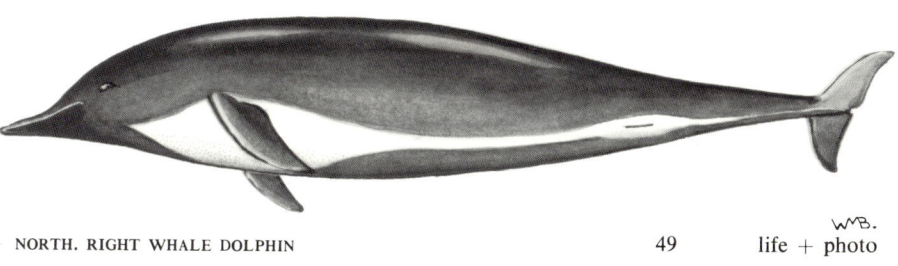

NORTH. RIGHT WHALE DOLPHIN   49   WMB.
                                        life + photo

STH. RIGHT WHALE DOLPHIN   50   WMB.
                                           photo

ROUGH TOOTHED DOLPHIN   51   WMB.
                                        life + illust.

ELLIOT'S DOLPHIN   52   life + illust.

WᴍB.

AMAZONE RIVER DOLPHIN   53   description + foto

WᴍB.

GUIANA RIVER DOLPHIN   54   frozen spec.

WᴍB.

CAMERUN RIVER DOLPHIN   55   photo

WᴍB.

LEADCOLOURED DOLPHIN   56   life + photo

SPECKLED DOLPHIN   57   life

BORNEO WHITE DOLPHIN   58   life + photo

CHINESE WHITE DOLPHIN   59   life + descr.

BOTTLENOSED DOLPHIN   60   life + photo

BOTTLENOSED DOLPHIN   60a   life + photo

BOTTLENOSED DOLPHIN   60b   life + photo

PACIFIC BOTTLENOSED DOLPHIN   61   life + photo

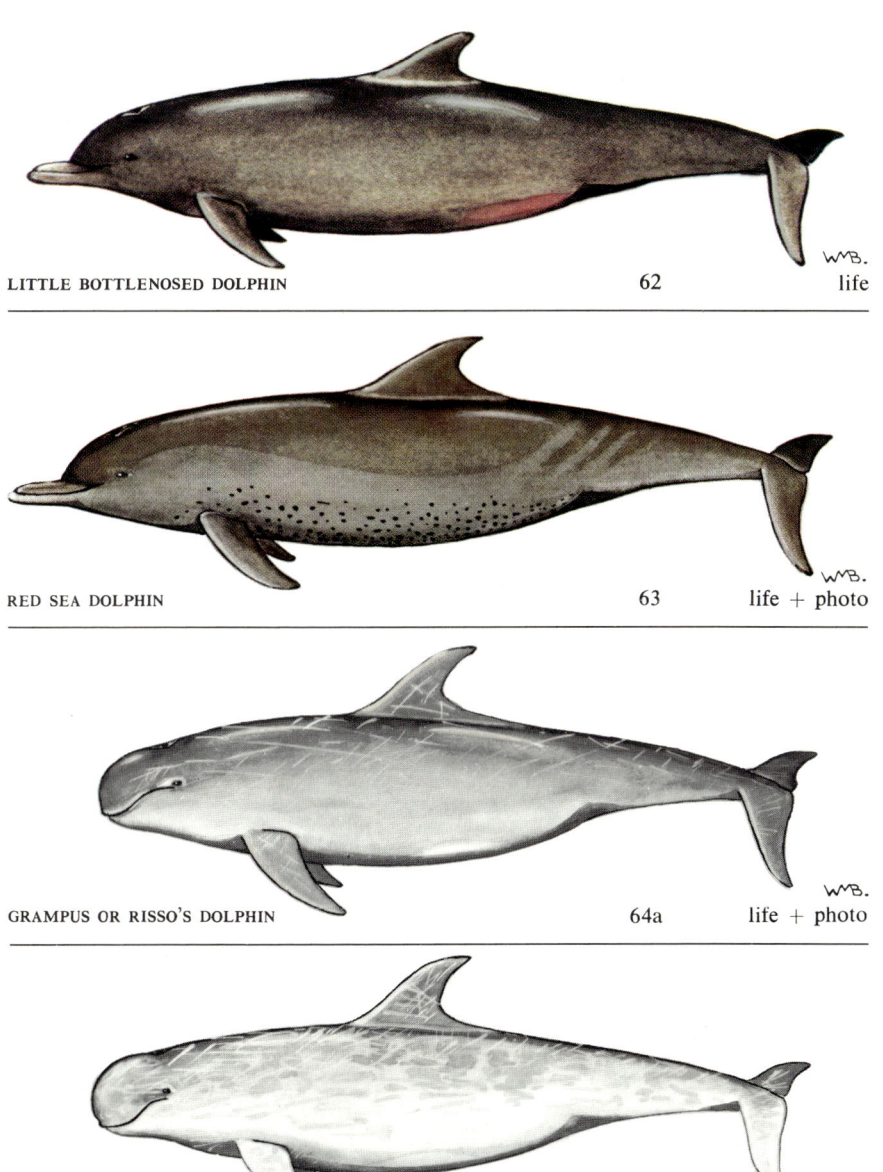

LITTLE BOTTLENOSED DOLPHIN   62   life

RED SEA DOLPHIN   63   life + photo

GRAMPUS OR RISSO'S DOLPHIN   64a   life + photo

GRAMPUS OR RISSO'S DOLPHIN   64b   life + photo

PILOT WHALE 65 W^B.
life + photo

FALSE KILLER WHALE 66 W^B.
life + photo

LITTLE KILLER 67 W^B.
photo + life + descr.

PYGMY KILLER   68   photo + illustr.

KILLER WHALE   69   life + photo

ALULA WHALE   70   life

IRRAWADI DOLPHIN   71   W͎B.
                                                                        life

TASMAN (BEAKED) WHALE   72   W͎B.
                                                                        photo

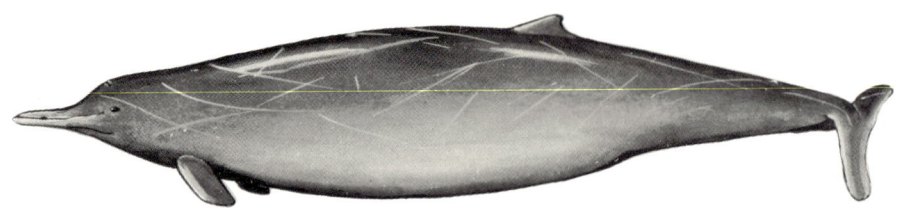

SOWERBY'S (BEAKED) WHALE   73   W͎B.
                                                                        illustr.

GULFSTREAM (BEAKED) WHALE   74   W͎B.
                                                                        illustr.

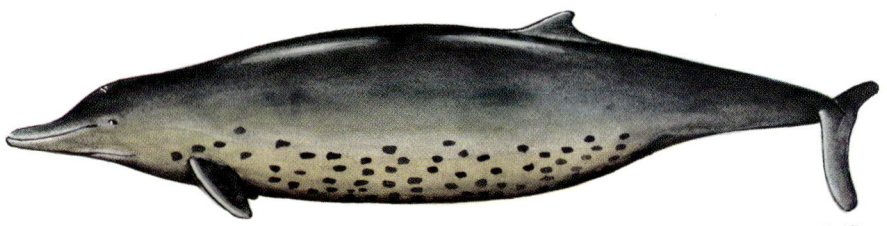

TRUE'S (BEAKED) WHALE   75   photo + description

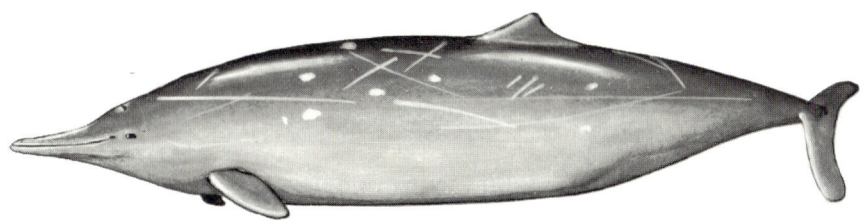

SCAMPERDOWN WHALE   76   life + descr.

SABRE-TOOTHED WHALE   77   photo

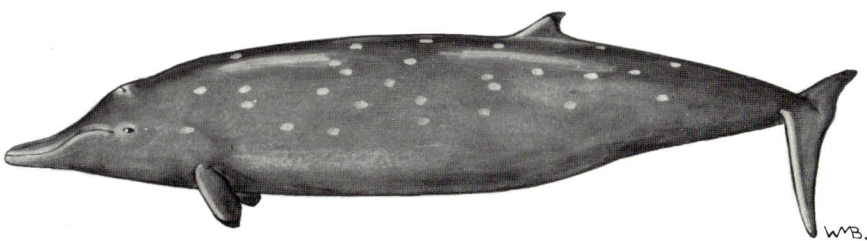

GINKGO WHALE   78   photo

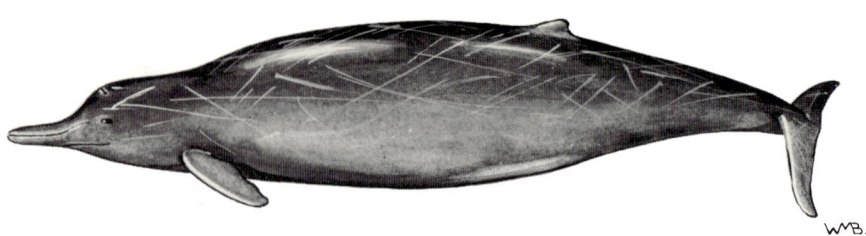

STRAPTOOTHED WHALE   79   photo

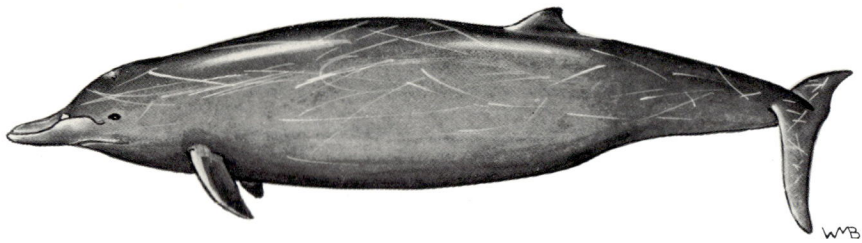

BLAINVILLE'S (BEAKED) WHALE   80   description + foto

CUVIER'S WHALE   81   life + photo

ARNOUX' WHALE   82   photo

BAIRD'S WHALE 83 WMB. photo

BOTTLENOSED WHALE 84 life + WMB. photo

SPERM WHALE 85 life + WMB. photo

PYGMY SPERM WHALE 86 life + WMB. photo

BLACK RIGHT WHALE   87   WᴍB.
photo + life

GREENLAND WHALE   88   WᴍB.
description

PYGMY RIGHT WHALE   89   WᴍB.
description

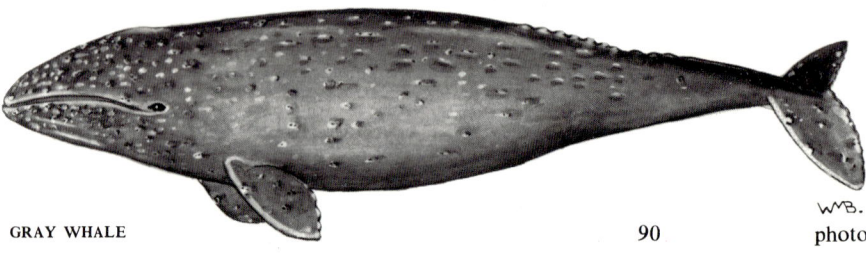

GRAY WHALE   90   WᴍB.
photo

MINKE WHALE   91   life + photo

SEI WHALE   92   life + photo

BRYDE'S WHALE   93   life + descr.

FIN WHALE   94   life + photo

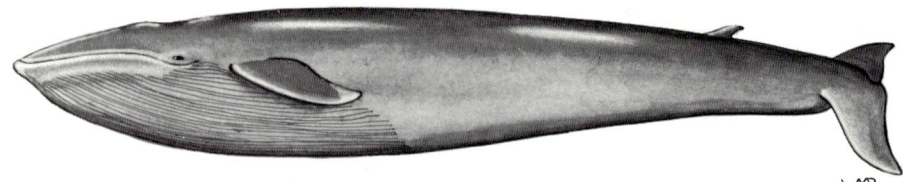

BLUE WHALE   95   life + illust.

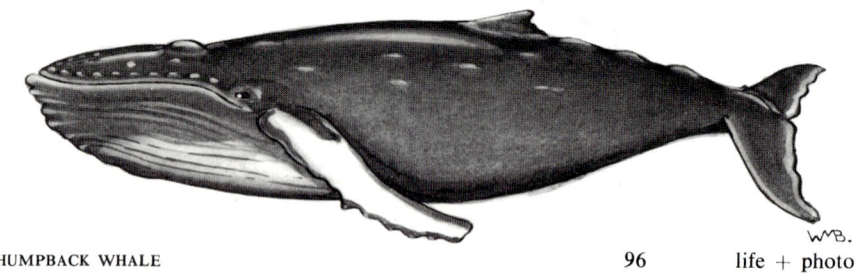

HUMPBACK WHALE   96   life + photo

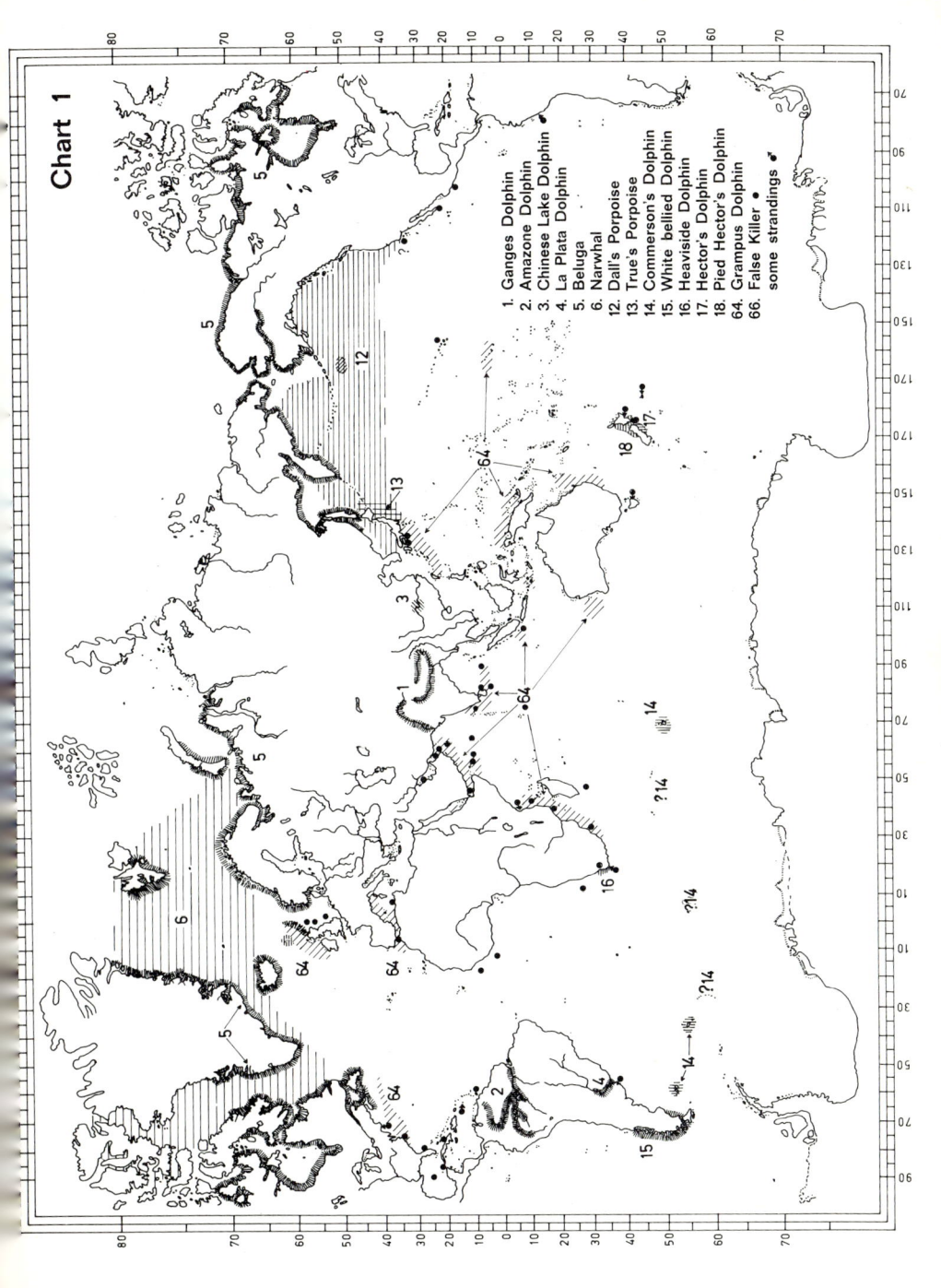

Chart 1

1. Ganges Dolphin
2. Amazone Dolphin
3. Chinese Lake Dolphin
4. La Plata Dolphin
5. Beluga
6. Narwhal
12. Dall's Porpoise
13. True's Porpoise
14. Commerson's Dolphin
15. White bellied Dolphin
16. Heaviside Dolphin
17. Hector's Dolphin
18. Pied Hector's Dolphin
64. Grampus Dolphin
66. False Killer ●
    some strandings ✐

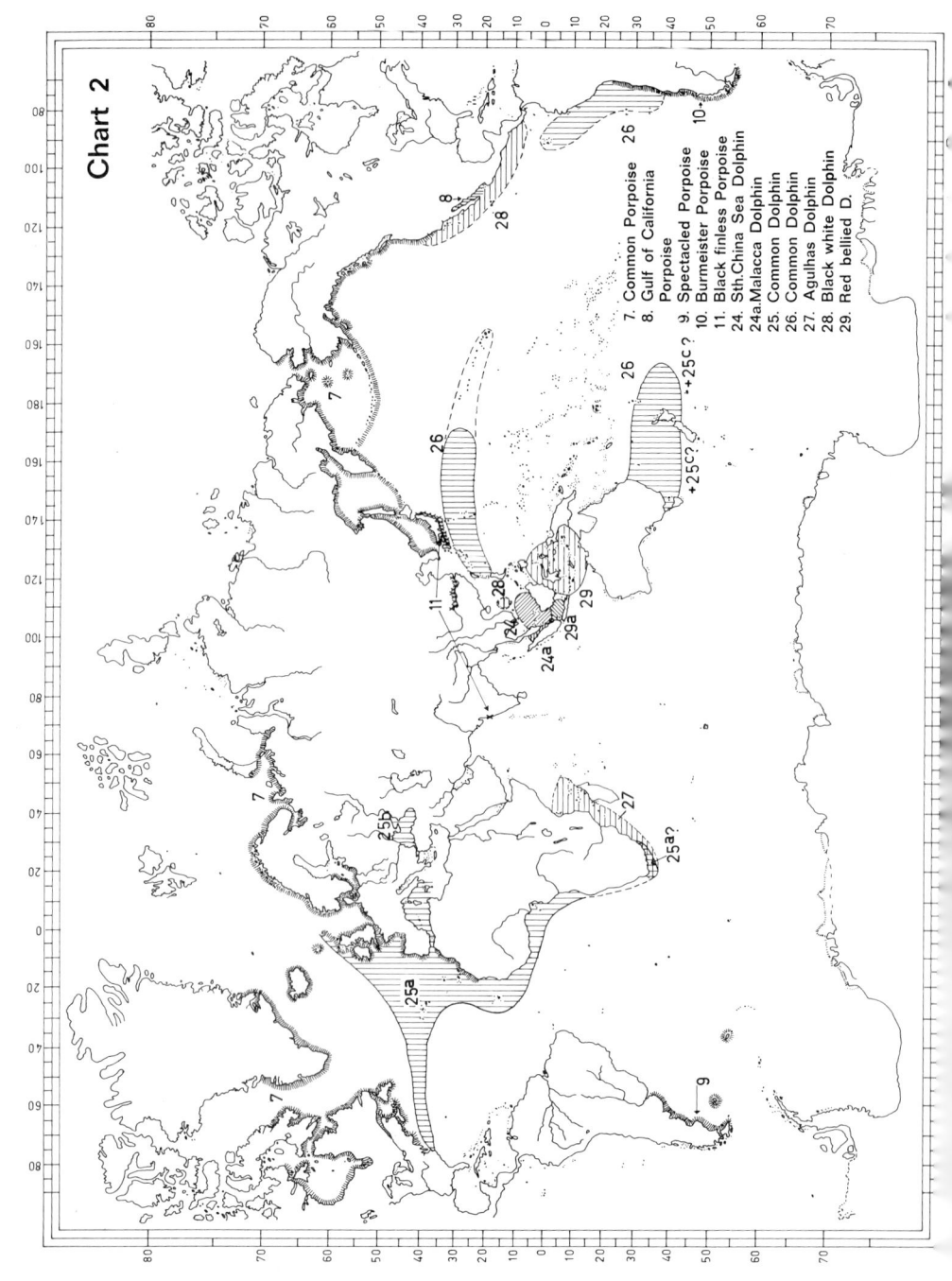

Chart 2

7. Common Porpoise
8. Gulf of California Porpoise
9. Spectacled Porpoise
10. Burmeister Porpoise
11. Black finless Porpoise
24. Sth.China Sea Dolphin
24a. Malacca Dolphin
25. Common Dolphin
26. Common Dolphin
27. Agulhas Dolphin
28. Black white Dolphin
29. Red bellied D.

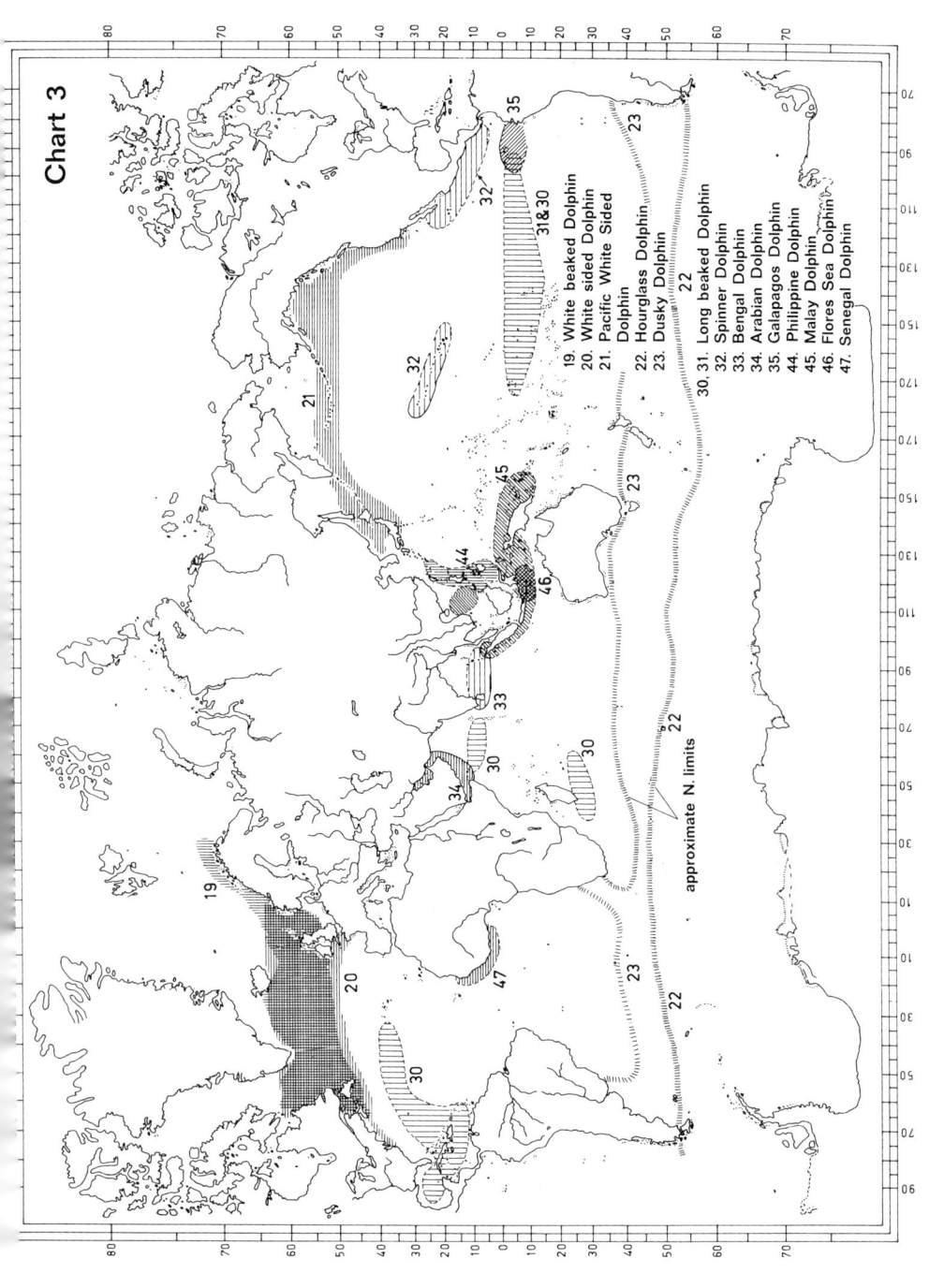

Chart 3

19. White beaked Dolphin
20. White sided Dolphin
21. Pacific White Sided
    Dolphin
22. Hourglass Dolphin
23. Dusky Dolphin

30, 31. Long beaked Dolphin
32. Spinner Dolphin
33. Bengal Dolphin
34. Arabian Dolphin
35. Galapagos Dolphin
44. Philippine Dolphin
45. Malay Dolphin
46. Flores Sea Dolphin
47. Senegal Dolphin

approximate N. limits

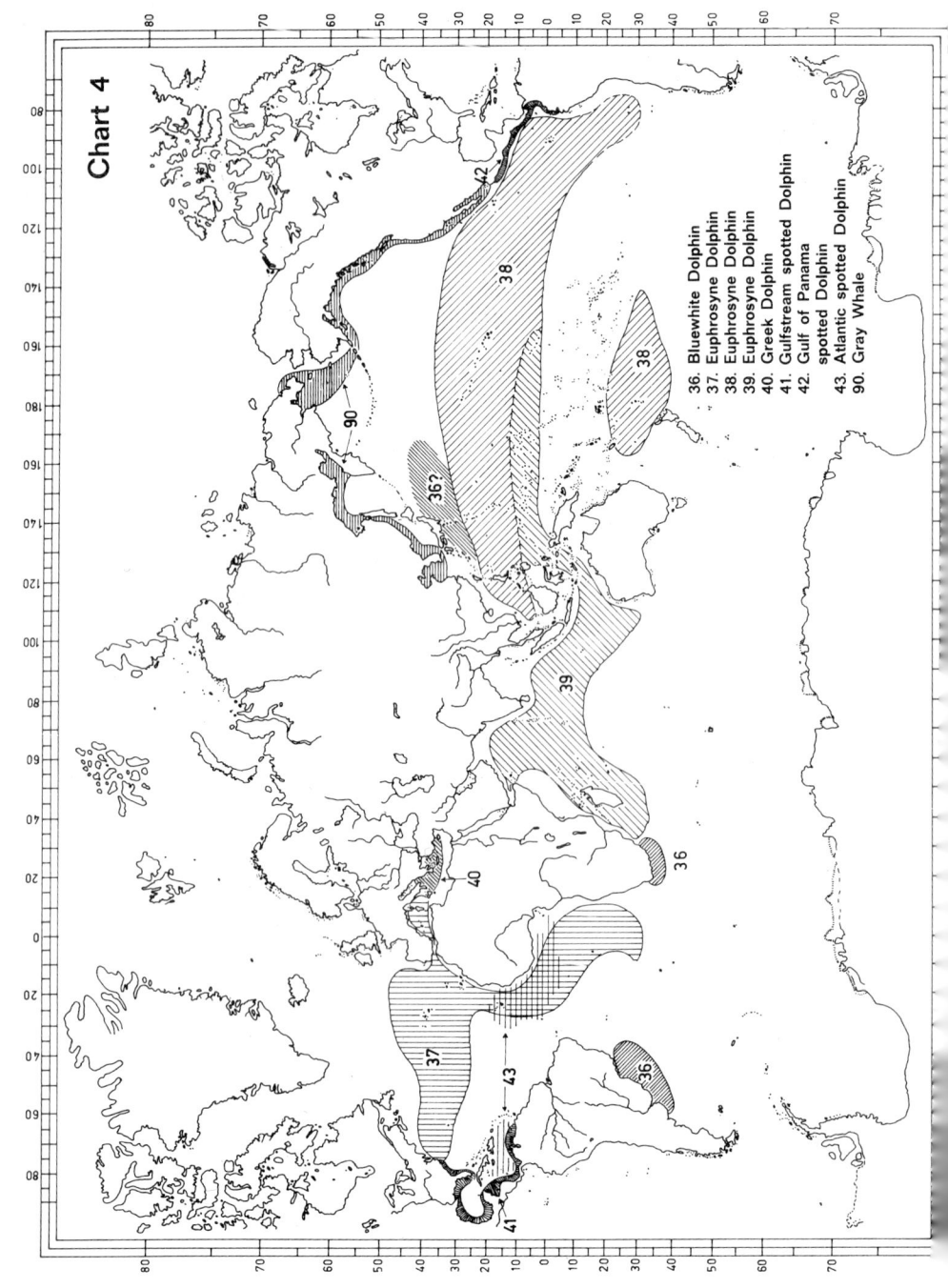

Chart 4

36. Bluewhite Dolphin
37. Euphrosyne Dolphin
38. Euphrosyne Dolphin
39. Euphrosyne Dolphin
40. Greek Dolphin
41. Gulfstream spotted Dolphin
42. Gulf of Panama
    spotted Dolphin
43. Atlantic spotted Dolphin
90. Gray Whale

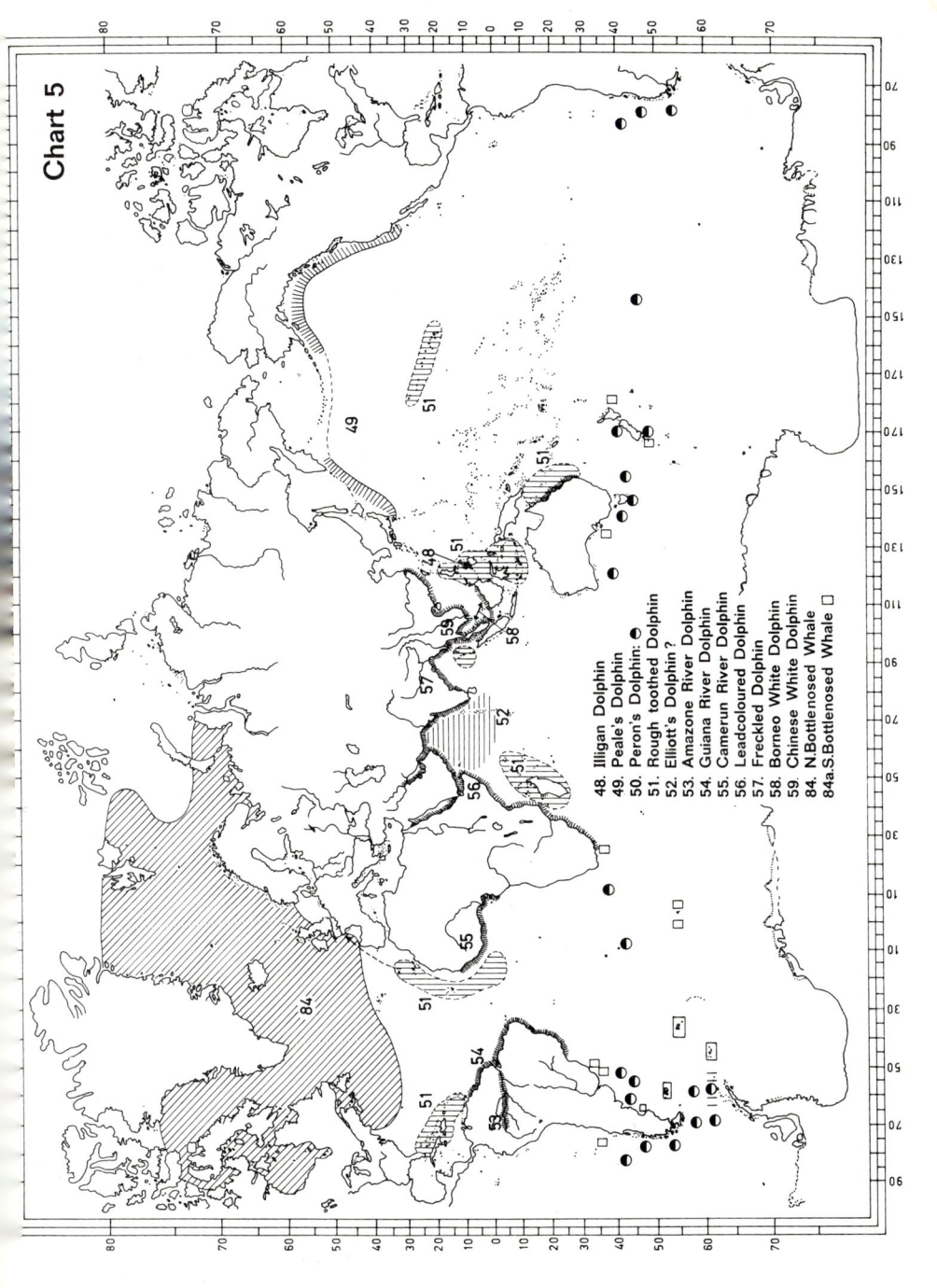

Chart 5

48. Illigan Dolphin
49. Peale's Dolphin
50. Peron's Dolphin
51. Rough toothed Dolphin
52. Elliott's Dolphin ?
53. Amazone River Dolphin
54. Guiana River Dolphin
55. Camerun River Dolphin
56. Leadcoloured Dolphin
57. Freckled Dolphin
58. Borneo White Dolphin
59. Chinese White Dolphin
84. N.Bottlenosed Whale
84a.S.Bottlenosed Whale

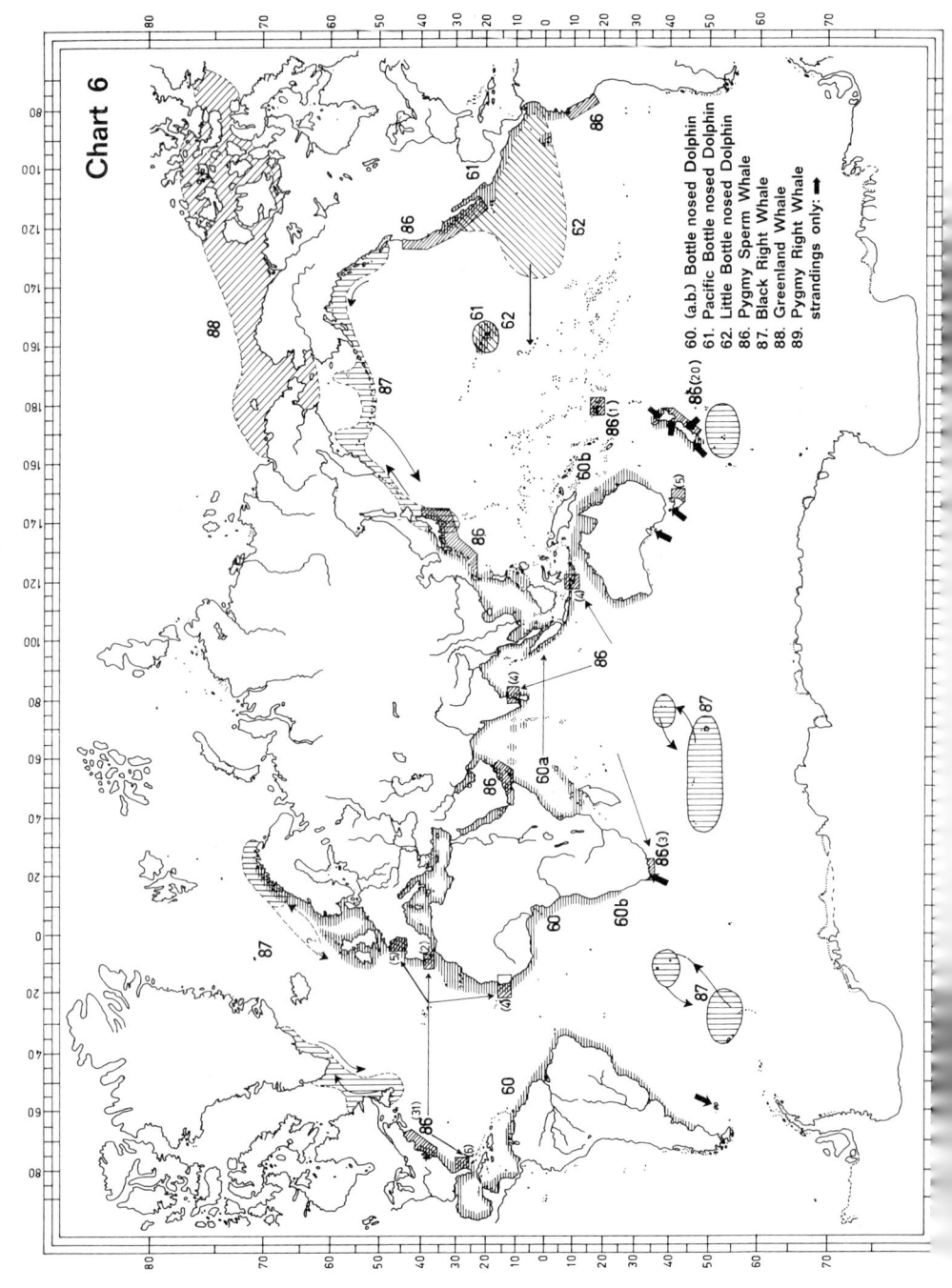

Chart 6

60. (a.b.) Bottle nosed Dolphin
61. Pacific Bottle nosed Dolphin
62. Little Bottle nosed Dolphin
86. Pygmy Sperm Whale
87. Black Right Whale
88. Greenland Whale
89. Pygmy Right Whale
    strandings only: →

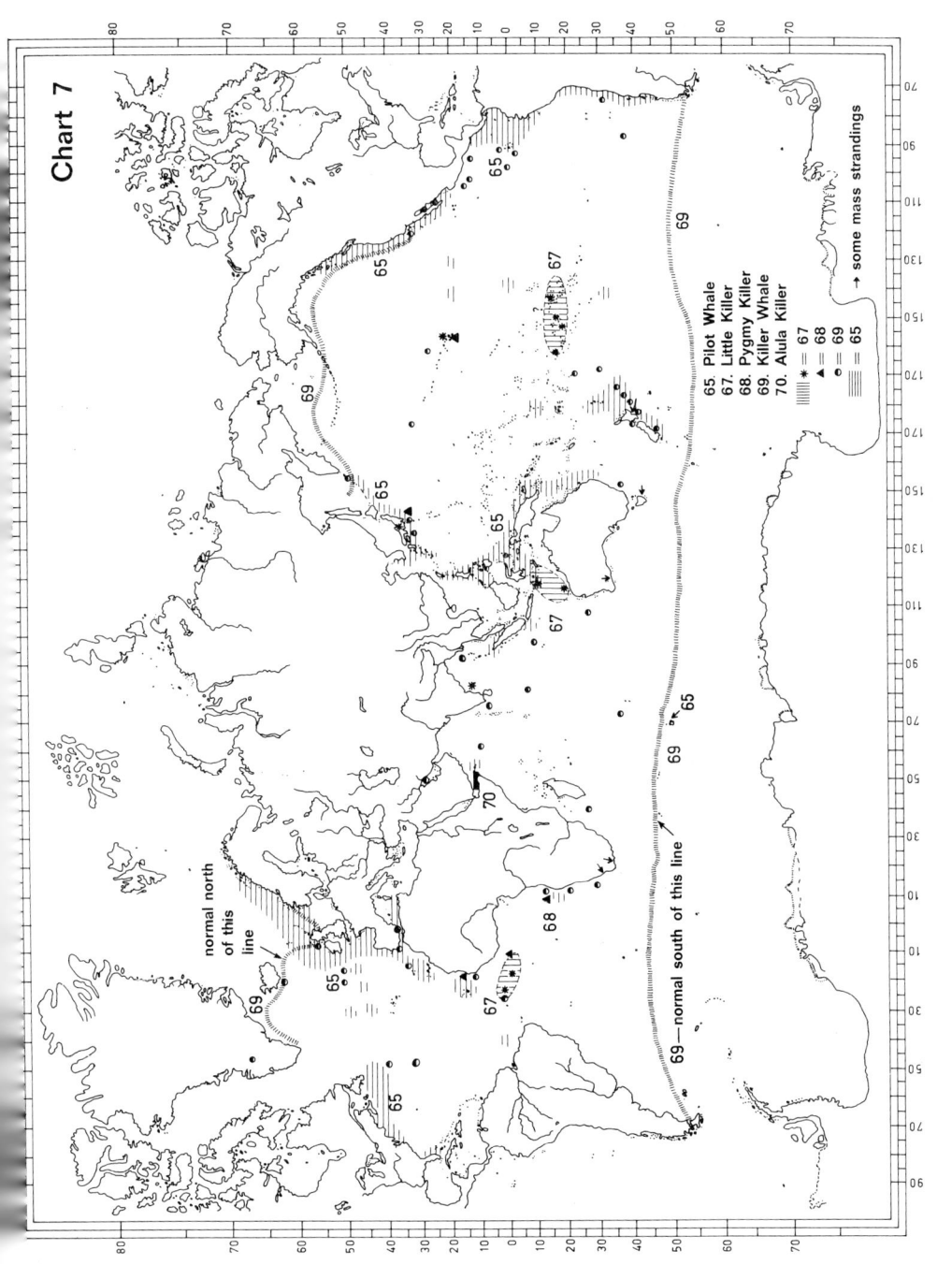

Chart 7

65. Pilot Whale
67. Little Killer
68. Pygmy Killer
69. Killer Whale
70. Alula Killer

|||||| = 67
||| = 68
||| = 69
||| = 65

→ some mass strandings

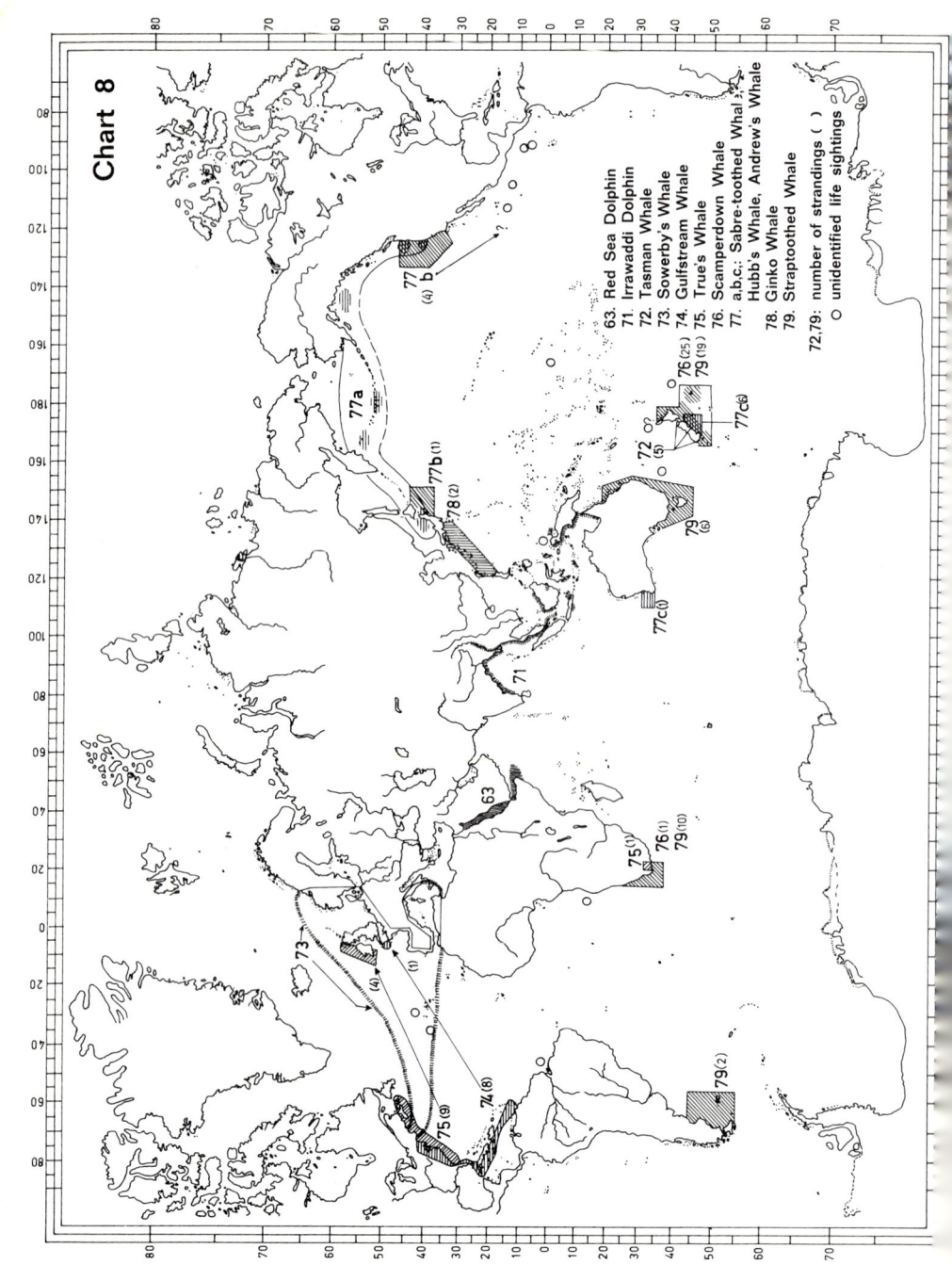

Chart 8

63. Red Sea Dolphin
71. Irrawaddi Dolphin
72. Tasman Whale
73. Sowerby's Whale
74. Gulfstream Whale
75. True's Whale
76. Scamperdown Whale
77. a,b,c,: Sabre-toothed Whale;
     Hubb's Whale, Andrew's Whale
78. Ginko Whale
79. Straptoothed Whale

72,79: number of strandings ( )
   ○ unidentified life sightings

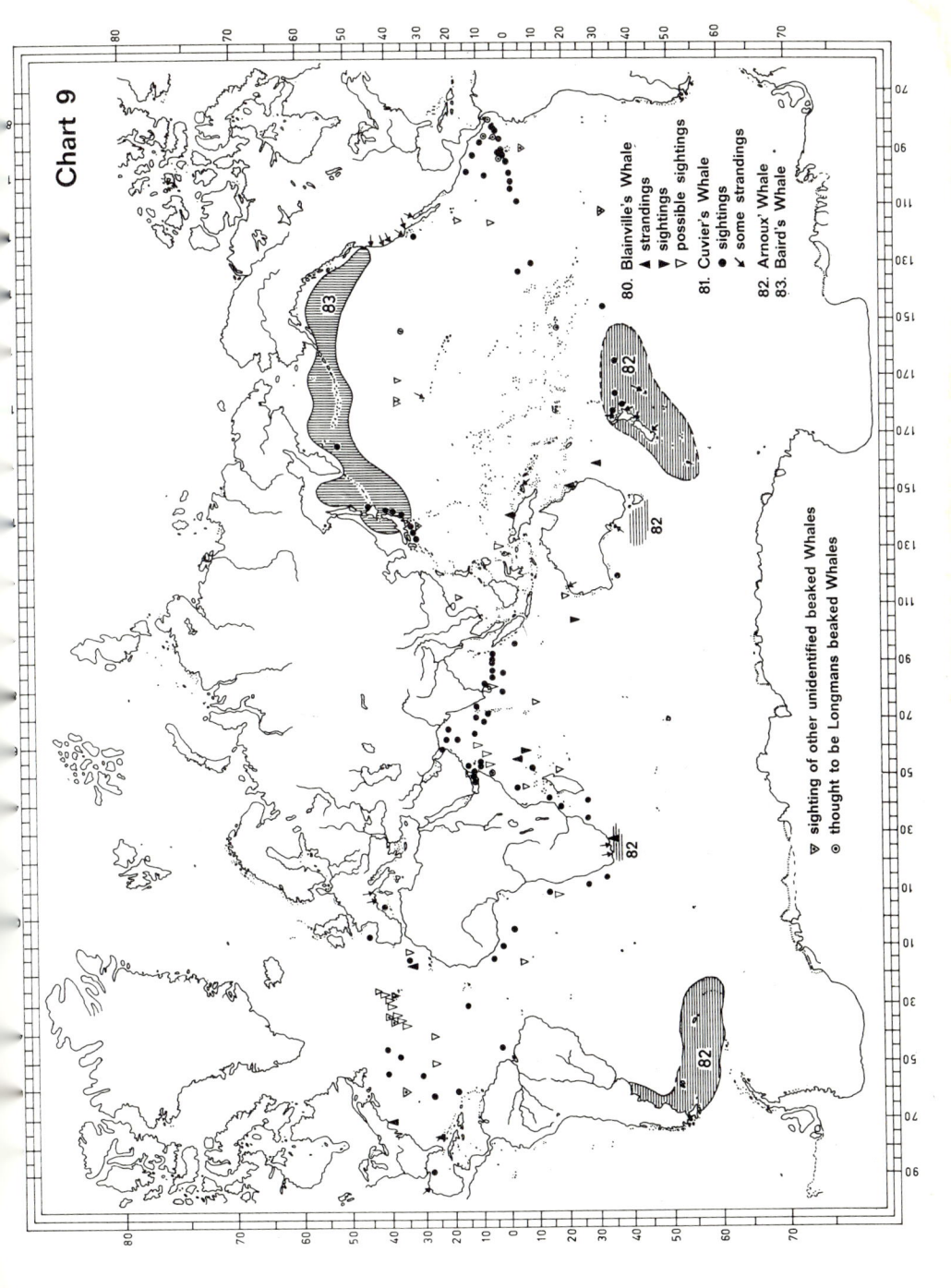

Chart 9

80. Blainville's Whale
    ▲ strandings
    ▼ sightings
    ▽ possible sightings
81. Cuvier's Whale
    ● sightings
    ⚓ some strandings
82. Arnoux' Whale
83. Baird's Whale

▽ sighting of other unidentified beaked Whales
⊙ thought to be Longmans beaked Whales

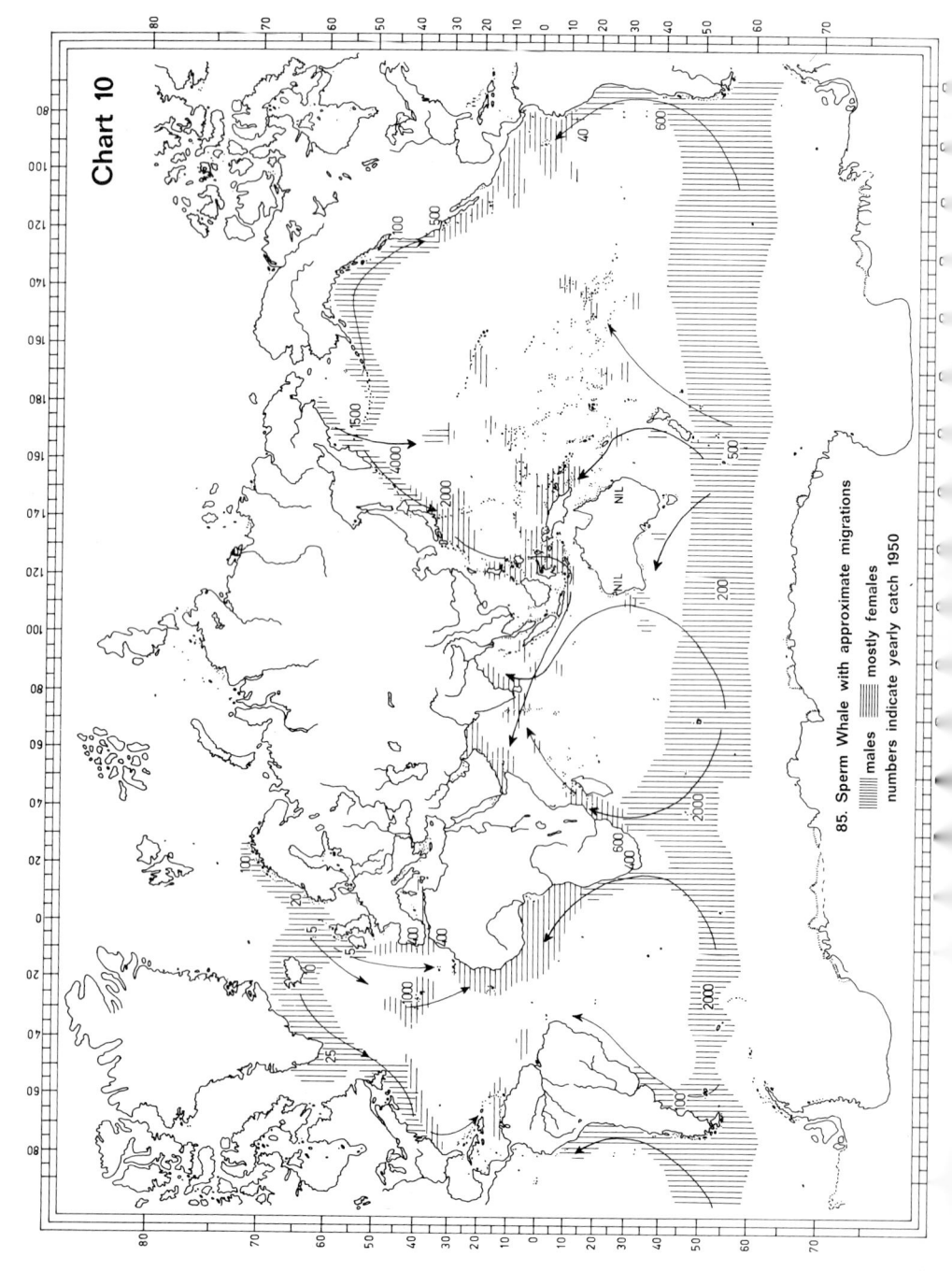

Chart 10

85. Sperm Whale with approximate migrations
males   mostly females
numbers indicate yearly catch 1950

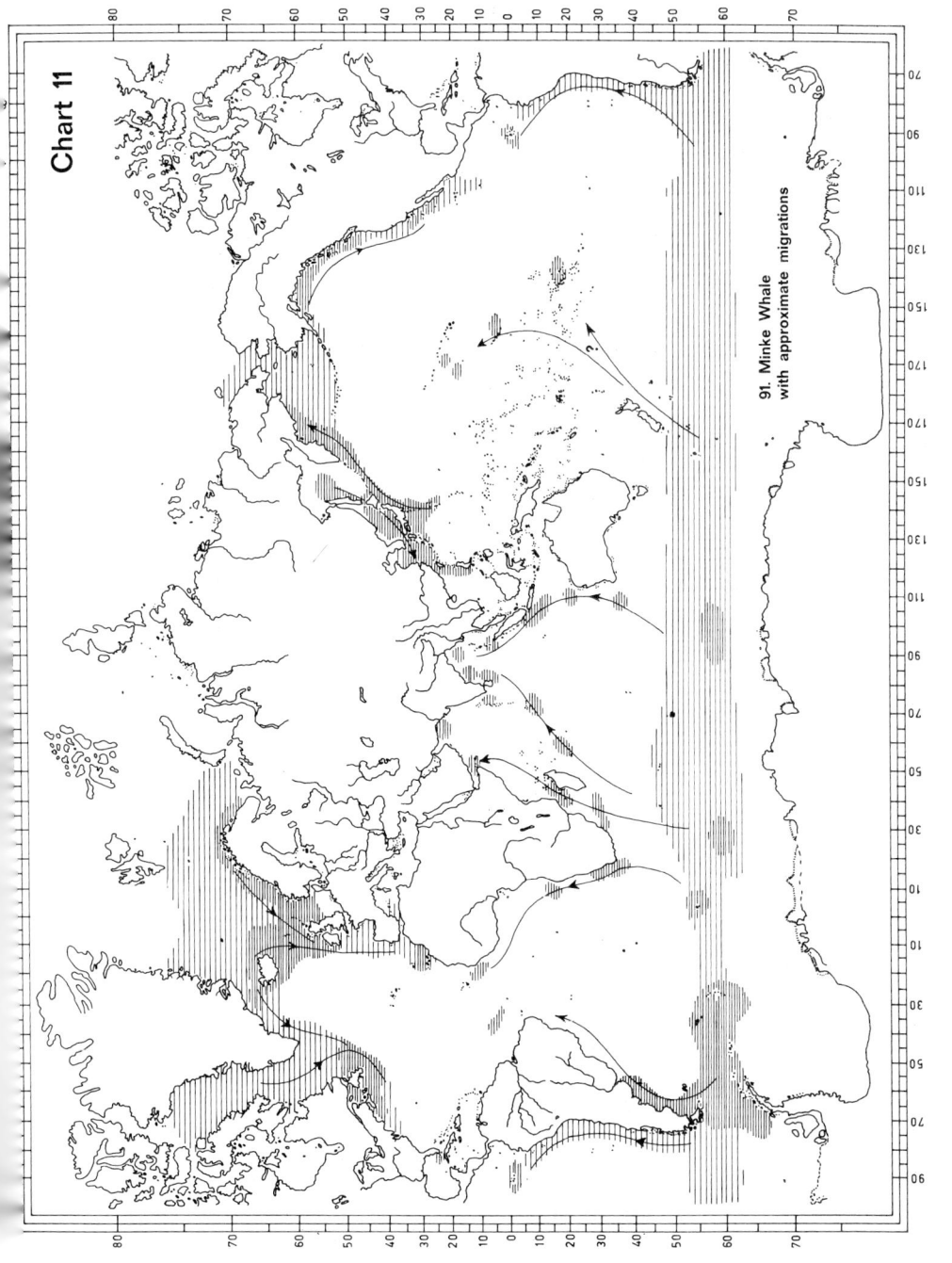

Chart 11

91. Minke Whale
with approximate migrations

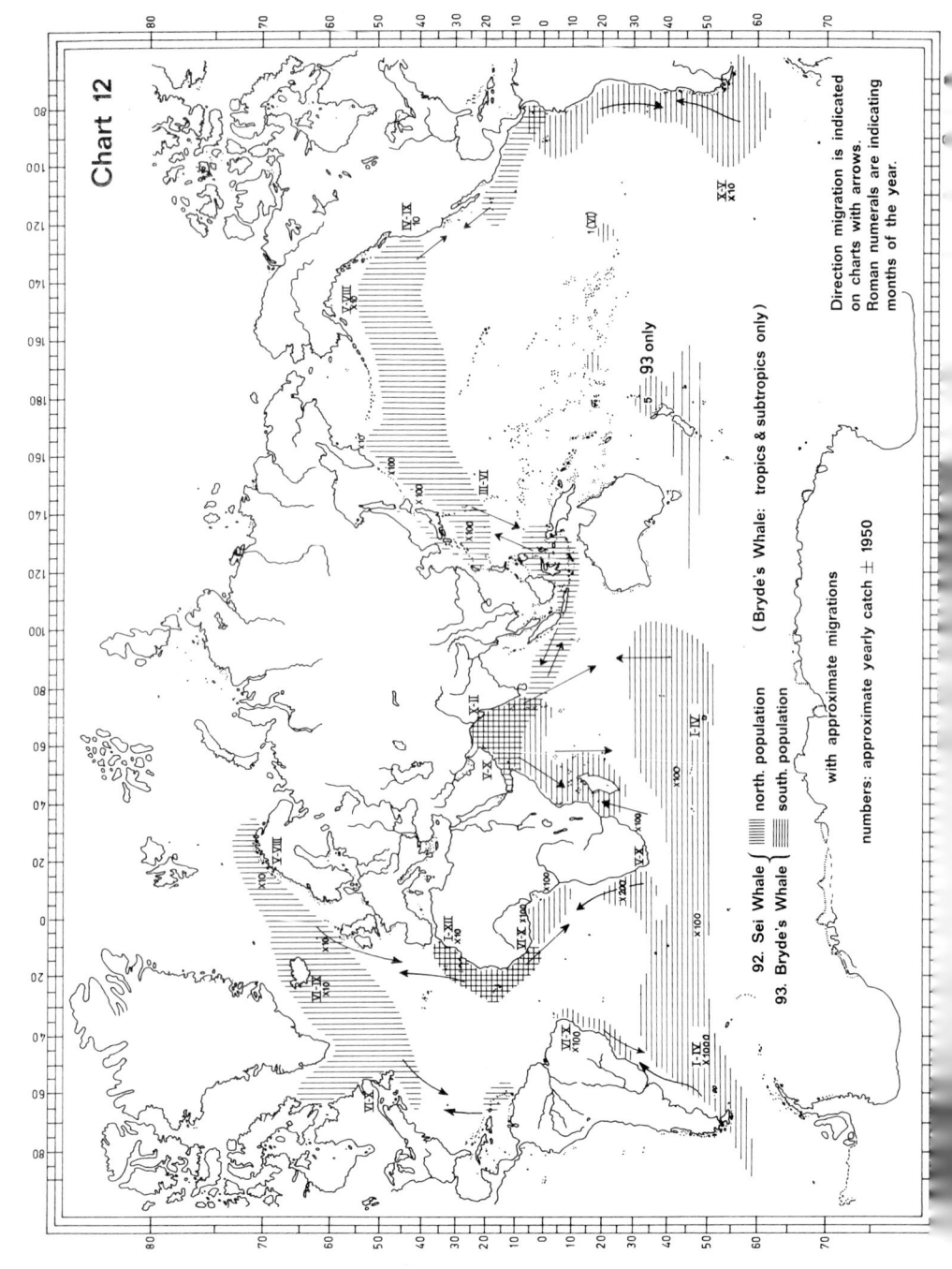

Chart 12

Direction migration is indicated on charts with arrows. Roman numerals are indicating months of the year.

(Bryde's Whale: tropics & subtropics only)

93 only

92. Sei Whale { north. population / south. population
93. Bryde's Whale { north. population / south. population

with approximate migrations

numbers: approximate yearly catch ± 1950

Chart 13

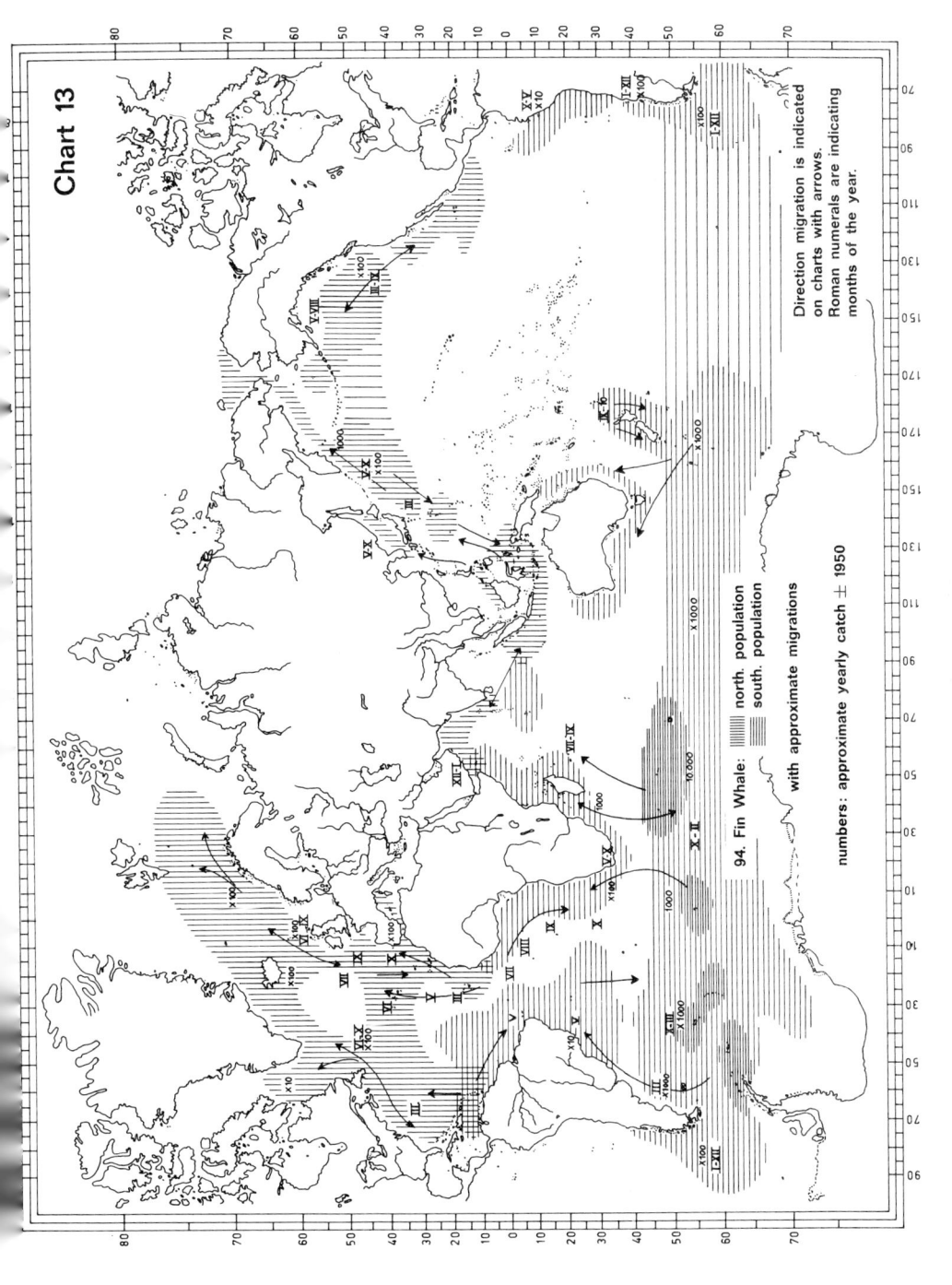

94. Fin Whale:
north. population
south. population
with approximate migrations

numbers: approximate yearly catch ± 1950

Direction migration is indicated
on charts with arrows.
Roman numerals are indicating
months of the year.

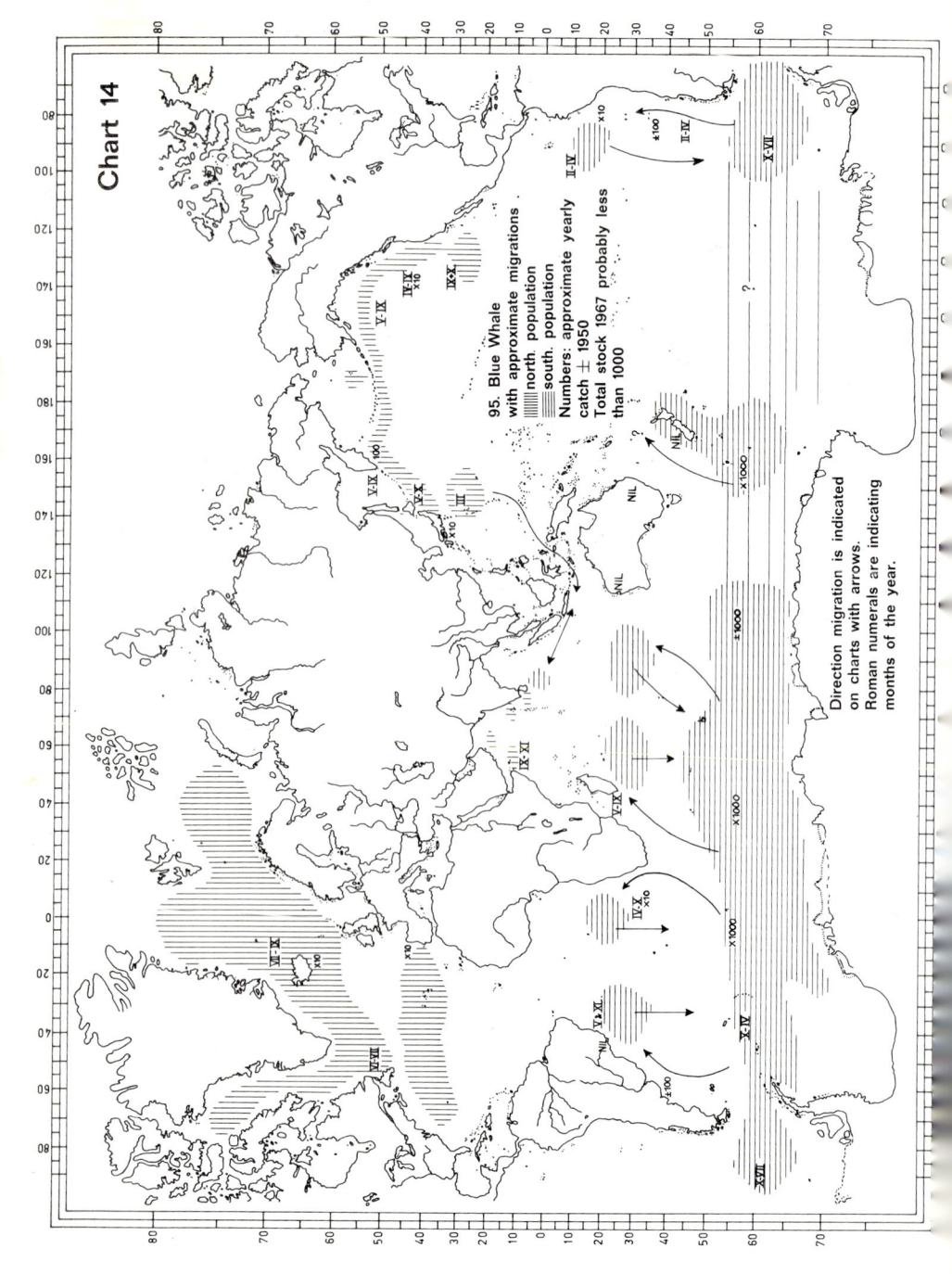

Chart 14

95. Blue Whale
with approximate migrations

▥ north. population
▤ south. population

Numbers: approximate yearly
catch ± 1950
Total stock 1967 probably less
than 1000

Direction migration is indicated
on charts with arrows.
Roman numerals are indicating
months of the year.

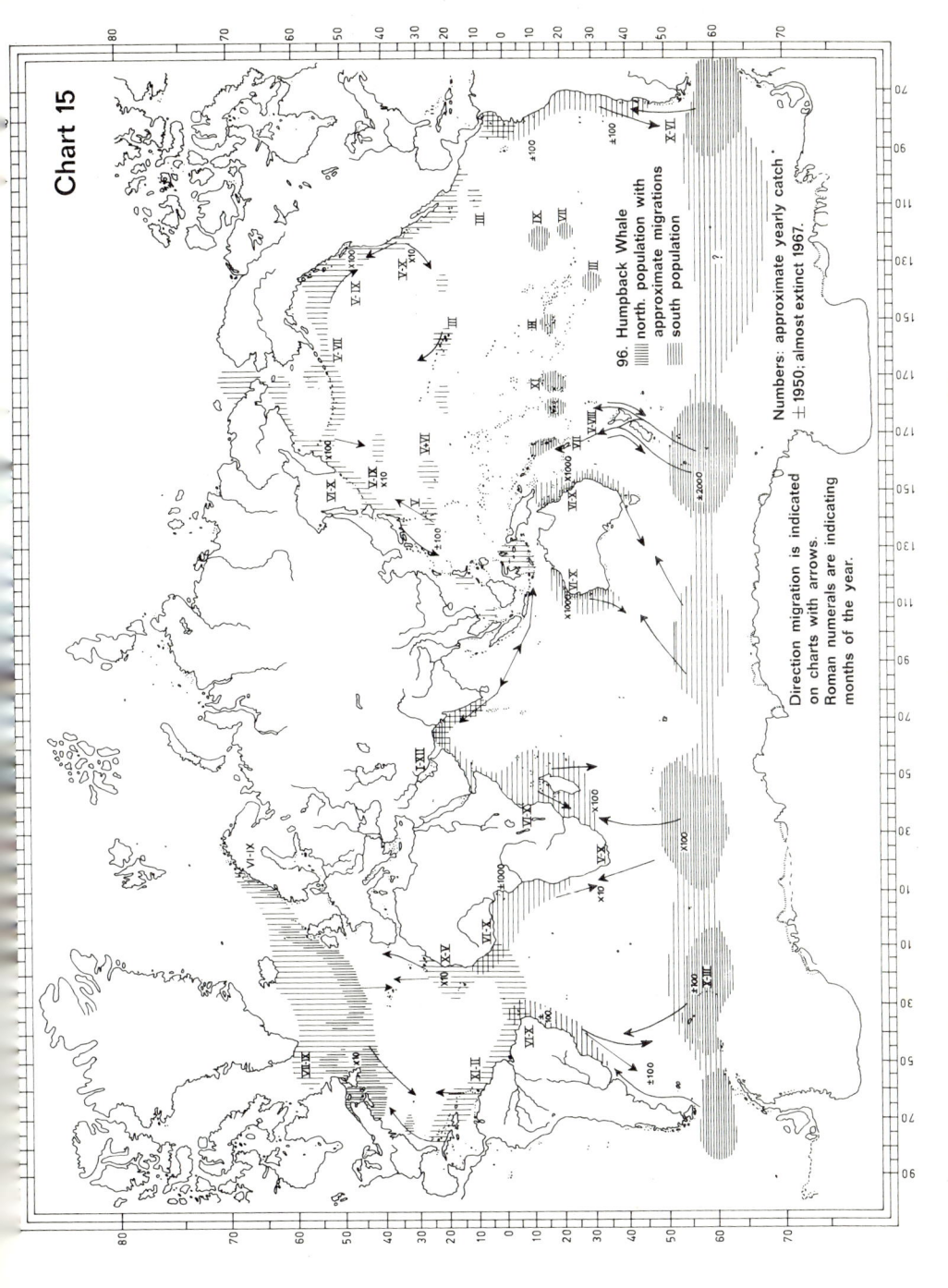

Chart 15

96. Humpback Whale
||||| north. population with
        approximate migrations
≡≡≡ south population

Numbers: approximate yearly catch
± 1950: almost extinct 1967.

Direction migration is indicated
on charts with arrows.
Roman numerals are indicating
months of the year.

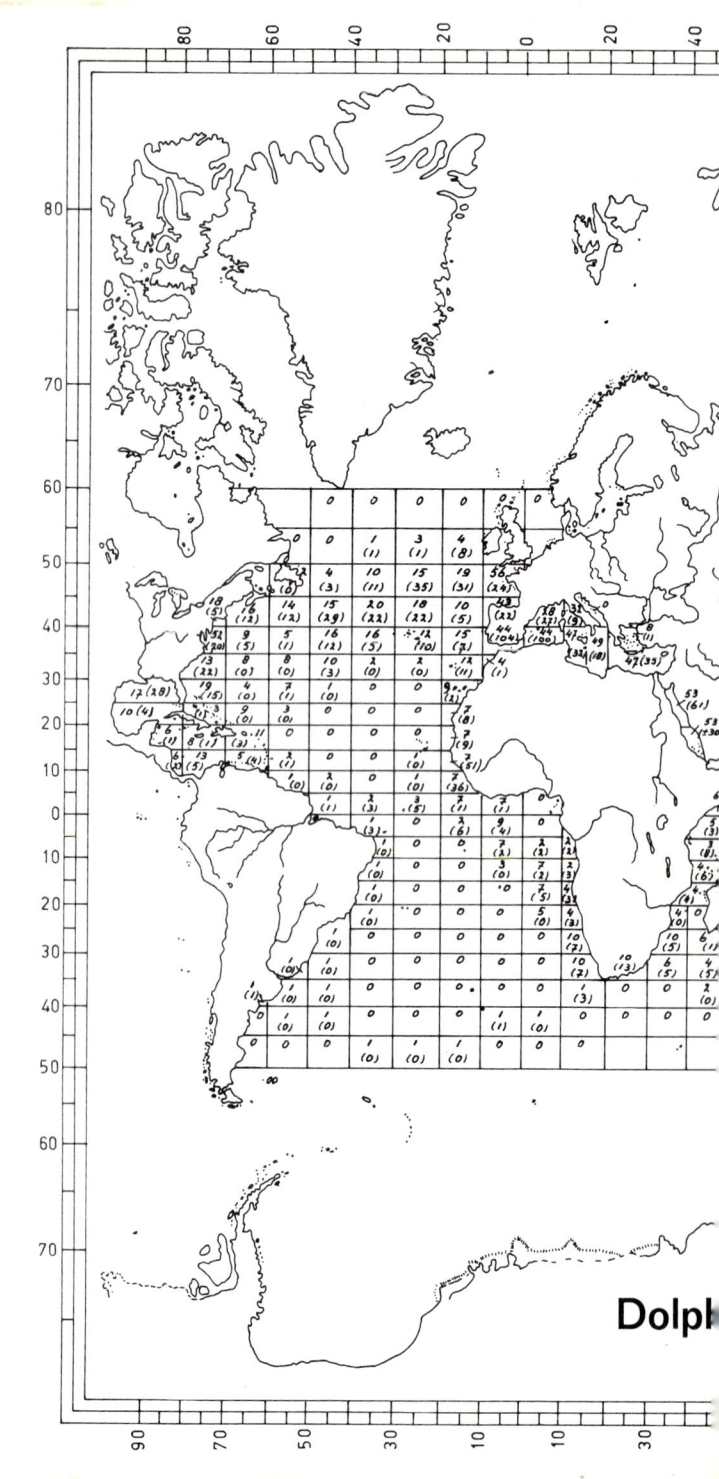

Dolph

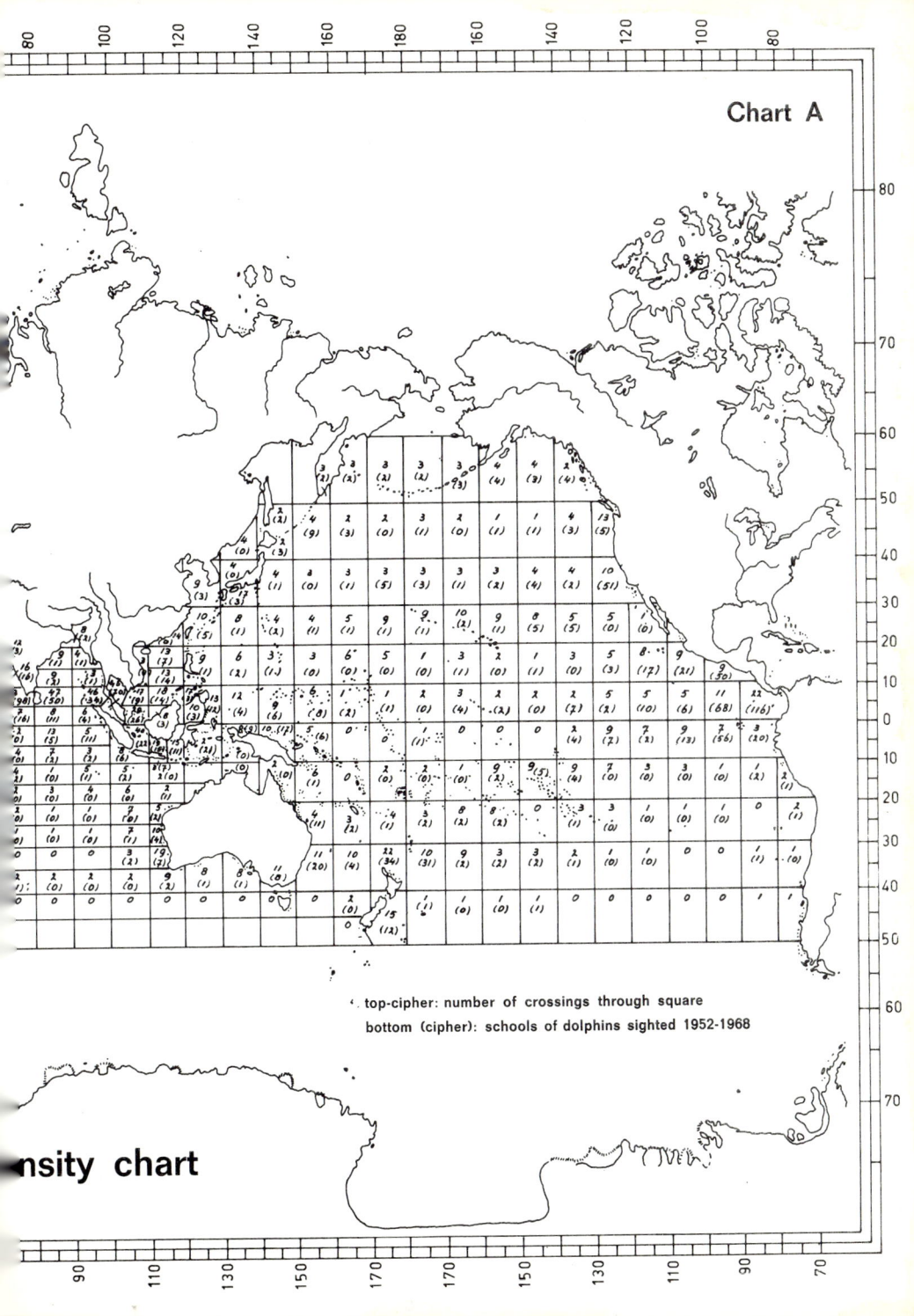

Chart A

top-cipher: number of crossings through square
bottom (cipher): schools of dolphins sighted 1952-1968

nsity chart

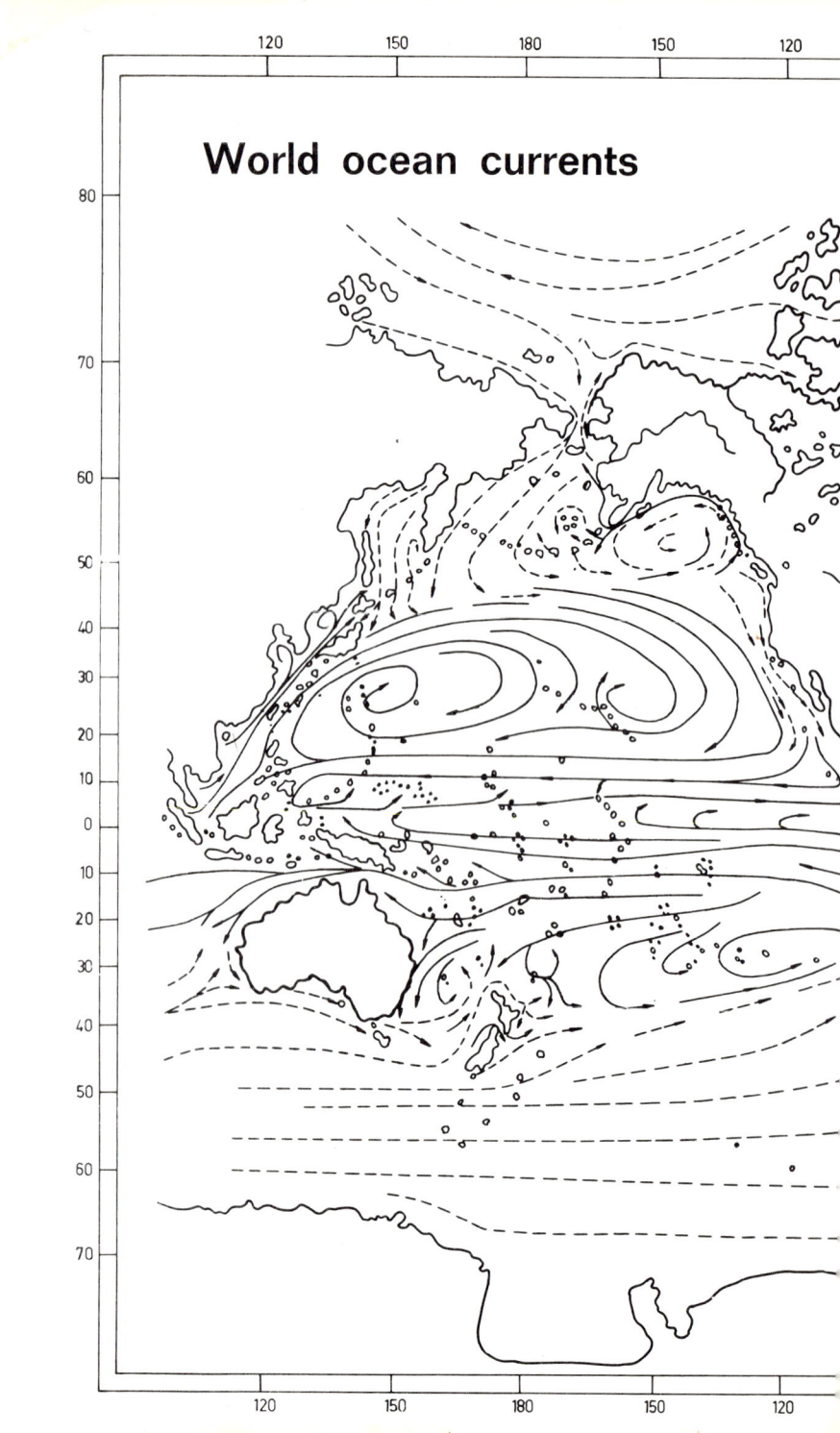

World ocean currents

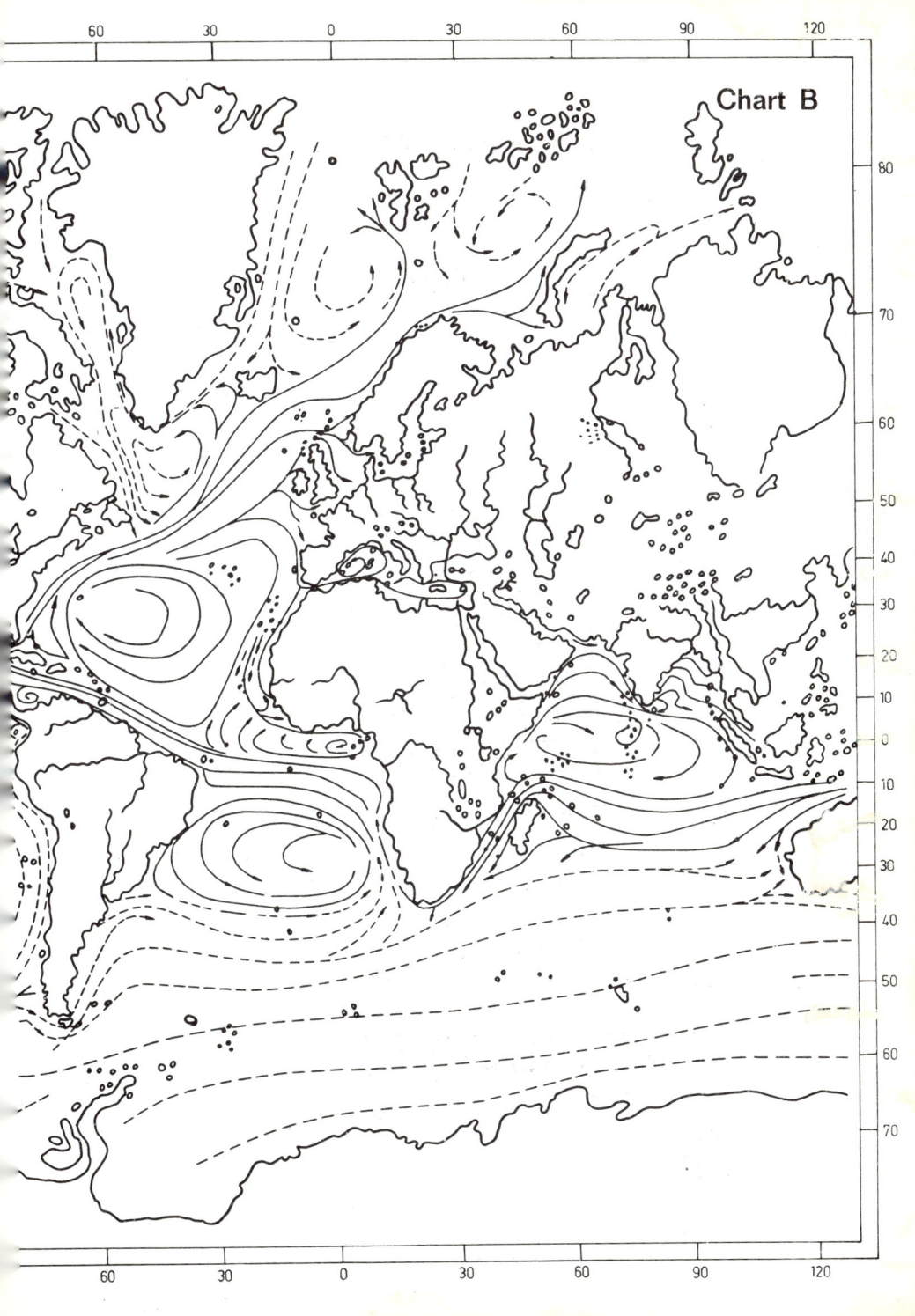

Chart B

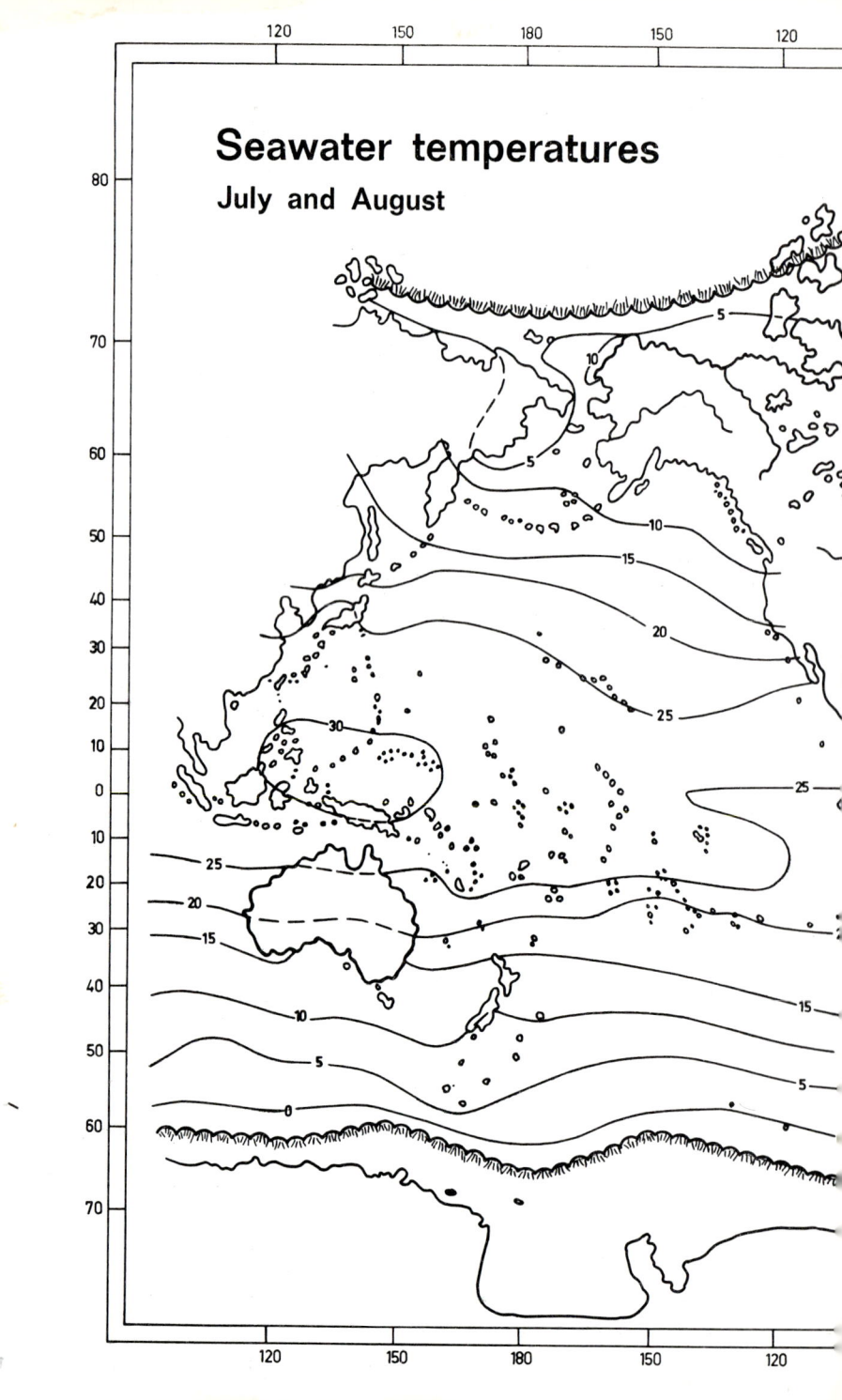

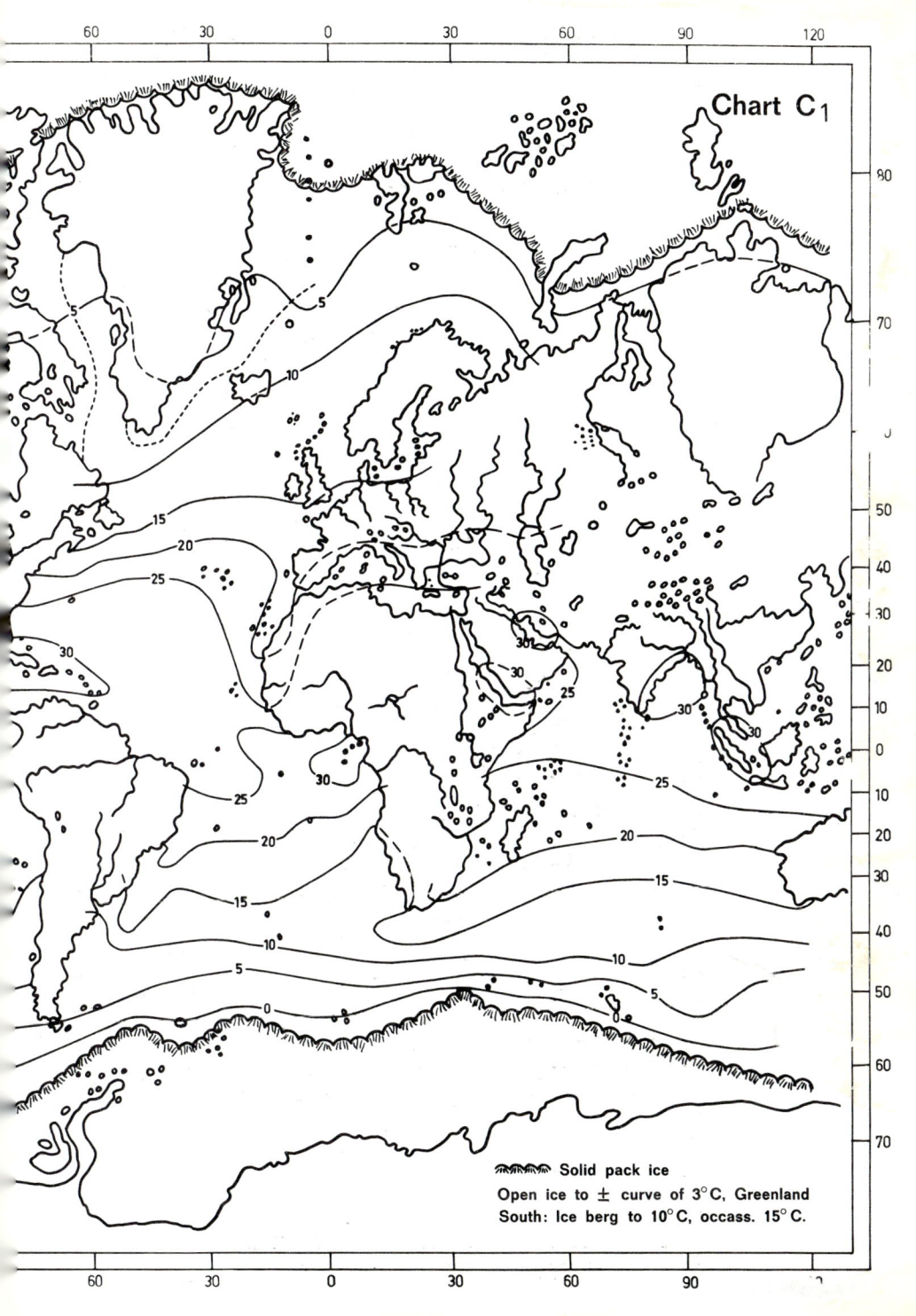

Chart C₁

Solid pack ice
Open ice to ± curve of 3°C, Greenland
South: Ice berg to 10°C, occass. 15°C.

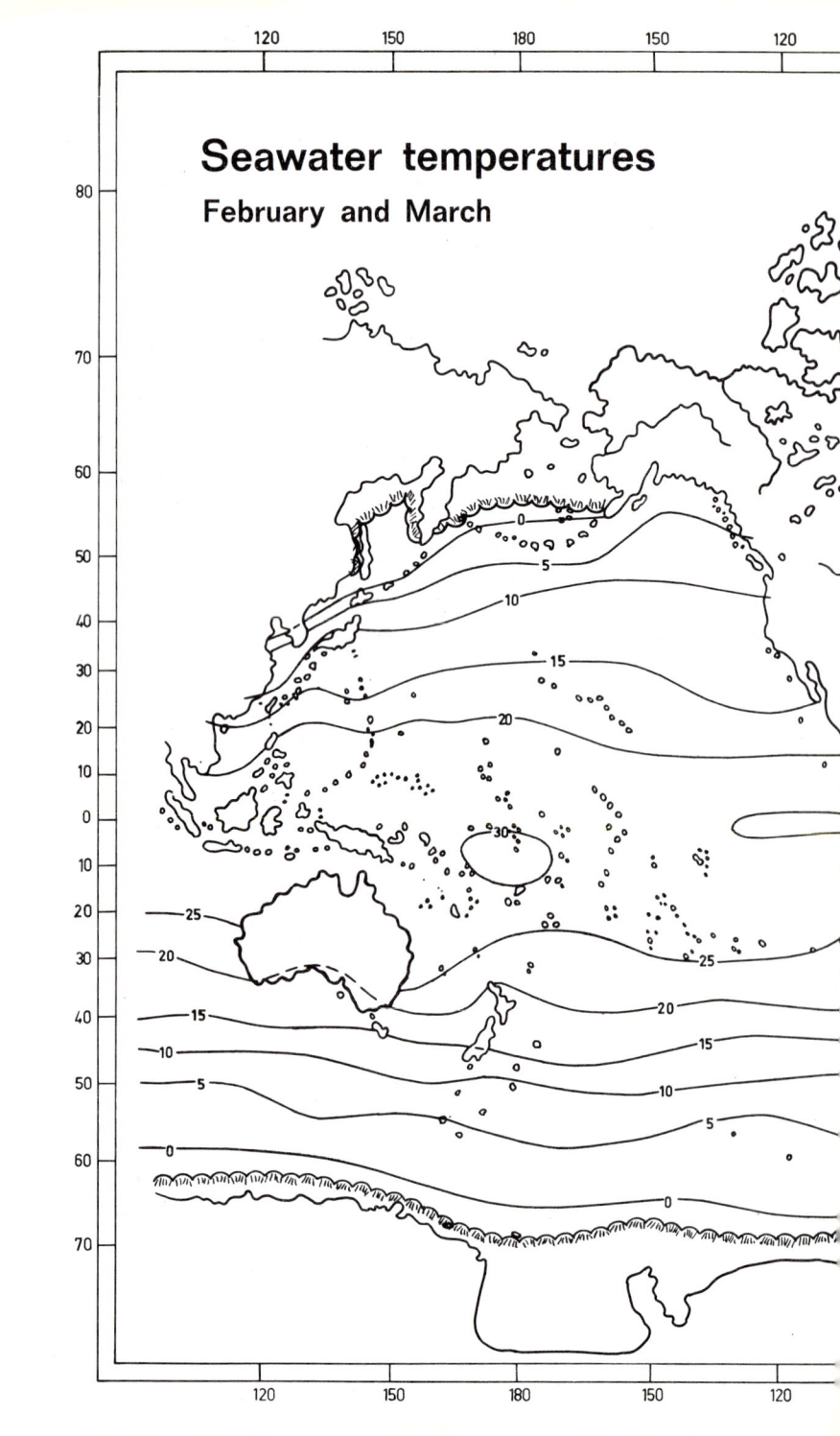

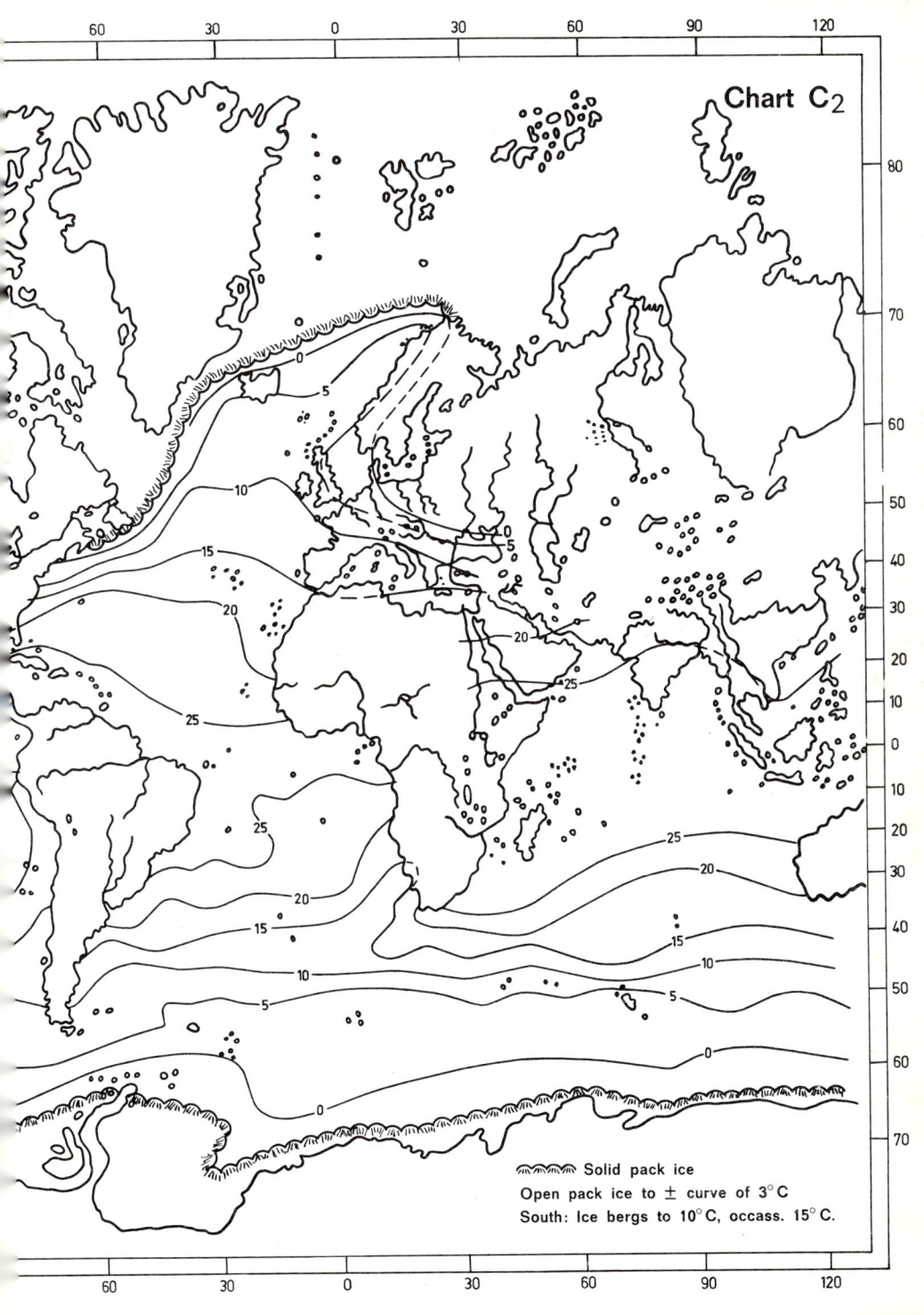

Chart C₂

Solid pack ice
Open pack ice to ± curve of 3°C.
South: Ice bergs to 10°C, occass. 15°C.

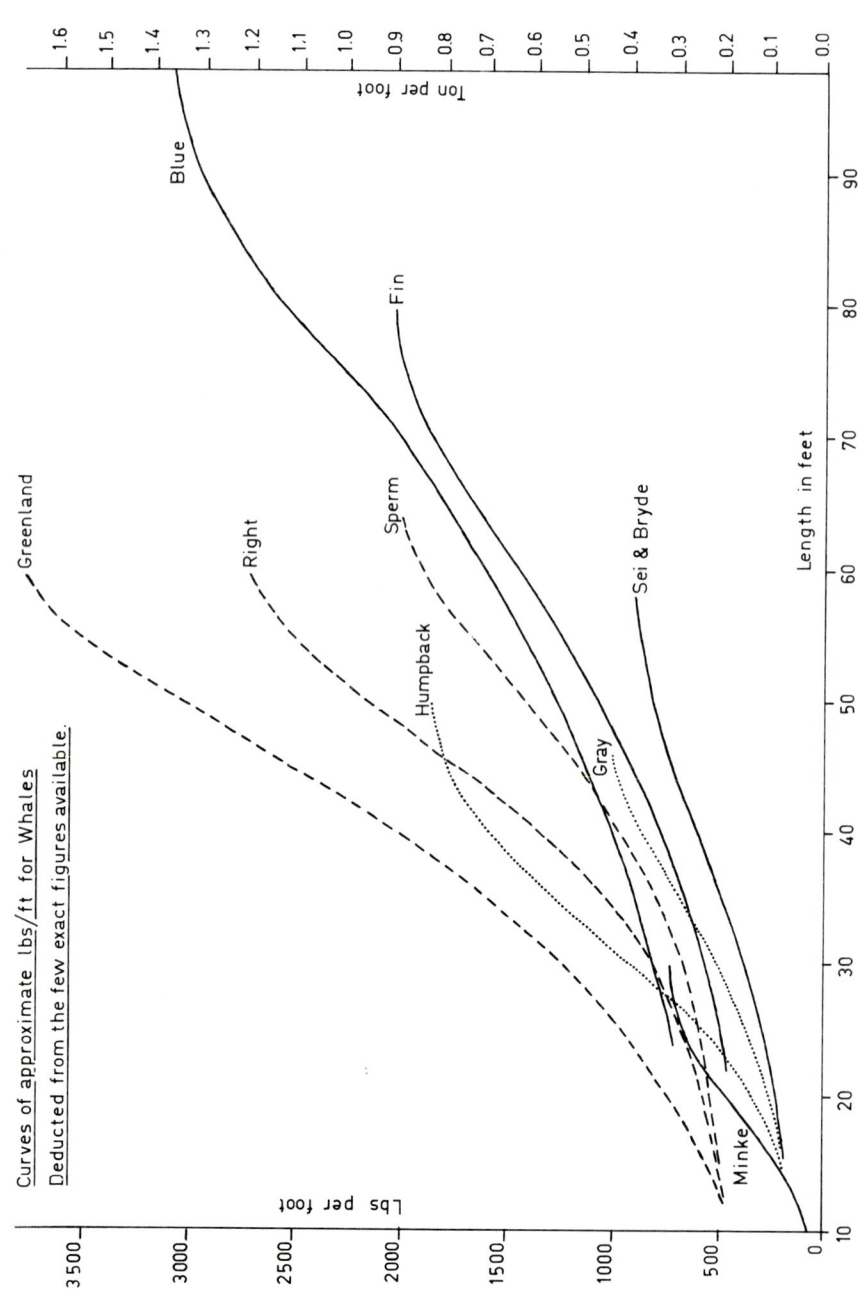

Curves of approximate lbs/ft for Whales

Deducted from the few exact figures available.

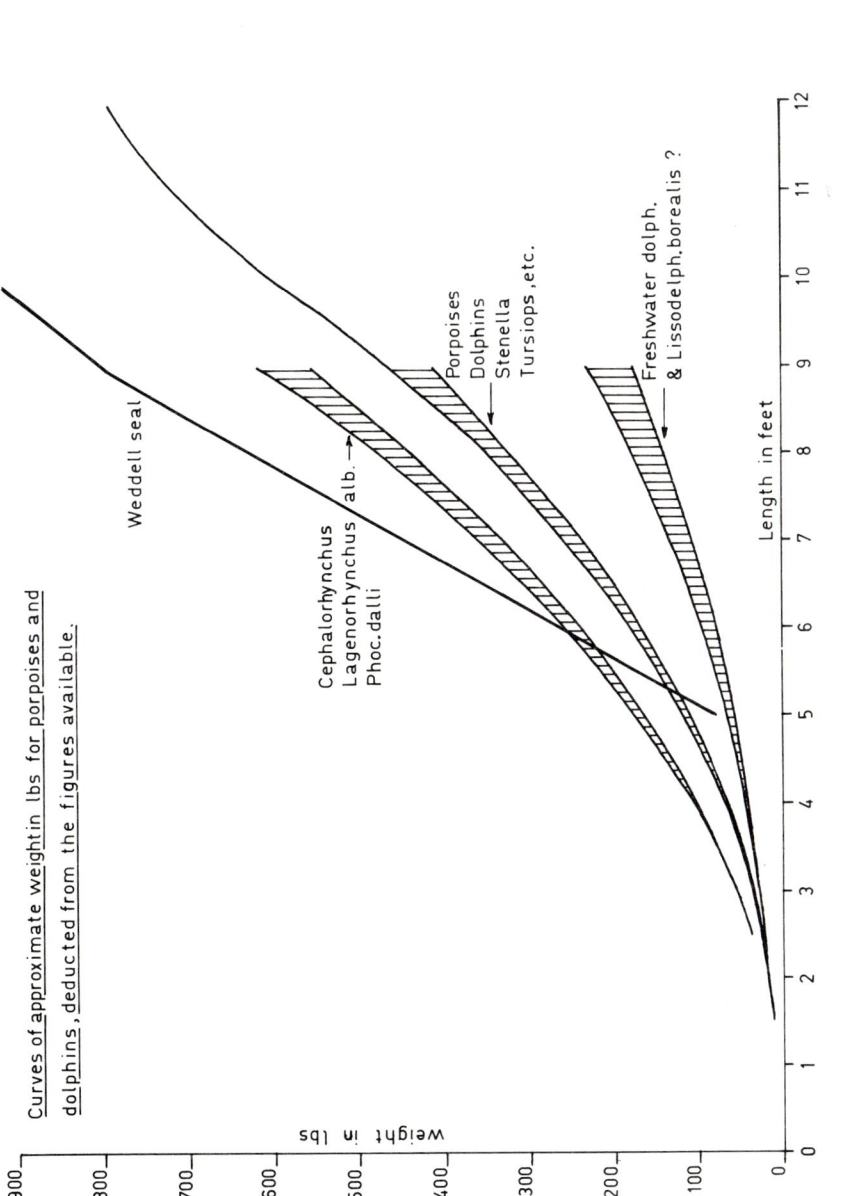

Curves of approximate weight in lbs for porpoises and dolphins, deducted from the figures available.

Weddell seal

Cephalorhynchus
Lagenorhynchus alb.
Phoc. dalli

Porpoises
Dolphins
Stenella
Tursiopps, etc.

Freshwater dolph.
& Lissodelph, borealis ?

Length in feet

weight in lbs

900
800
700
600
500
400
300
200
100
0

0 1 2 3 4 5 6 7 8 9 10 11 12

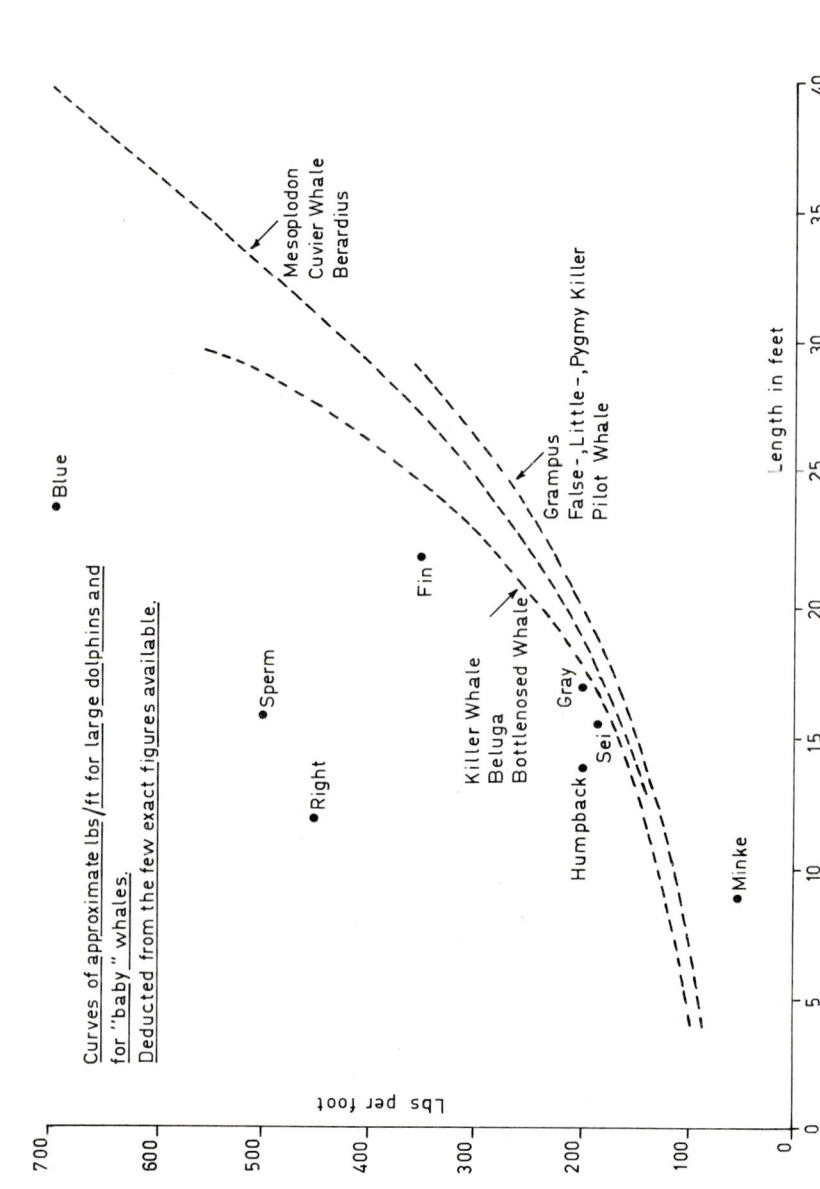

Curves of approximate lbs/ft for large dolphins and for "baby" whales.
Deducted from the few exact figures available.

Lbs per foot

700
600
500
400
300
200
100
0

5 10 15 20 25 30 35 40

Length in feet

• Blue
• Sperm
• Right
• Fin
• Minke

Killer Whale
Beluga
Bottlenosed Whale

Humpback •
Sei •
Gray •

Mesoplodon
Cuvier Whale
Berardius

Grampus
False-, Little-, Pygmy Killer
Pilot Whale

Contents

Colophon

Composed in the Times Roman

Cover   John Egbert Seubring
Lay-out Jan Emmink

Printed by G. Kolff & Co.
Illustrations and charts W. F. J. Mörzer Bruyns

ISBN 90 70055 090

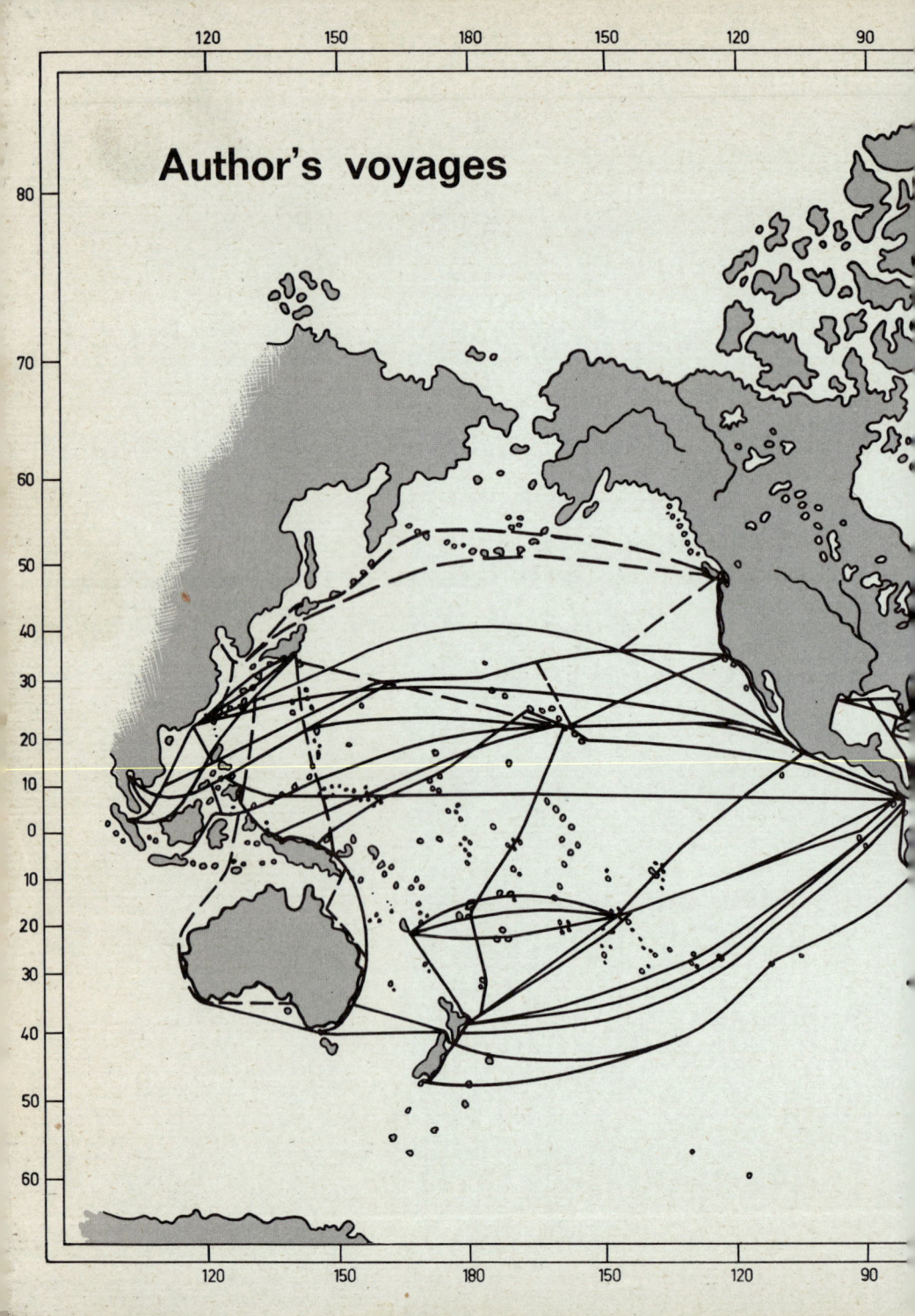

Author's voyages